No Sense of Obligation
Science and Religion in an Impersonal Universe

Matt Young

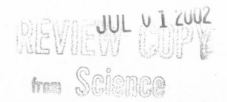

Recommended Cataloging Data
Library of Congress Control Number: BL240.2
Dewey Decimal System Call Number: 240.2

ISBN: 0-75961-089-4

The author gratefully acknowledges permission to reprint
previously published material from The TANAKH: The New
JPS Translation According to the Traditional Hebrew Text.
Copyright 1985 by the Jewish Publication Society. Used by
permission.

This book is printed on acid free paper.

Matt Young, PhD
Department of Physics
Colorado School of Mines
Golden, Colorado 80401

1stBooks – rev. 5/23/01

Praise for *No Sense of Obligation* by writers, scientists, clergy

Matt Young, a physicist by trade, provides a fascinating analysis of religious belief and argues that faith alone is not a valid reason for believing in anything whatever. In a day when there are influential religious leaders, such as Pat Robertson, preaching and sincerely believing that the entire vast physical universe is only about 7,500 years old, there is the most urgent need for books like *No Sense of Obligation*.

> STEVE ALLEN
> Author, composer, entertainer

No Sense of Obligation is a *tour de force* of science and religion, reason and faith, denoting in clear and unmistakable language and rhetoric what science really reveals about the cosmos, the world, and ourselves. Readers, both skeptics and believers, will be forced to confront their assumptions head-on and clarify their thinking on some of the deepest psychological and philosophical issues of our time and of all time. This book is, indeed, an antidote to fundamentalism, one of the most invidious ideological diseases of our culture.

> MICHAEL SHERMER
> Publisher, *Skeptic* Magazine
> Author, *How We Believe: The Search for God in an Age of Science*

Professor Young's book could not have come at a better time, a time when the perennial question of the relationship between religion and science has emerged again with particular urgency. The issues of religious fundamentalism, of sin, evil, and repentance have moved into the center of the arena of public discourse and public policy.

iii

At the same time there is a great interest in the truths and resources to be found in classic religious texts and the relative value and place of those set forth by scientific investigation.

With Young's unimpeachable foundation in the "hard sciences" and technology, the recognition of his colleagues, his obvious grounding in religious texts and thought, as well as his direct involvement in a religious community, he offers us an important resource for continued rational discussion of live issues: the grounding of our quest for truth and faith, the basis of religious authority, and the problem of evil.

HERBERT BRONSTEIN
Senior Scholar, North Shore Congregation Israel, Glencoe, Illinois
Editor, *A Passover Haggadah, Five Scrolls for the Jewish Festivals*

I was brought up on a farm in rural Michigan. We were Seventh-Day Adventists, and my mother used to read us the Bible every day. Eventually, I went to college and became a physicist. I was amazed that Matt and I have come to very similar conclusions, even though we began at opposite poles. His book makes the skeptic's case brilliantly. Whether you agree with him or not, it makes for interesting thought provoking reading, and I highly recommend it.

DONALD JENNINGS
Physicist (retired), National Institute of Standards and Technology

The author is a sympathetic yet forceful critic of orthodox religious belief. I cannot accept his conclusions, but this is a serious and thoughtful work and presents a challenge that believers should not simply dismiss.

Believers will find a comprehensive and fascinating range of topics, a precise summary of major arguments in opposition to

iv

accepted important religious doctrines and Scriptural interpretations. Having them in this form will help in being prepared to answer or respond to the author's point of view.

His well-arranged order and sharply argued presentation will stimulate believers to reexamine their usual response to this type of debate.

DANIEL GOLDBERGER
Rabbi, Congregation Beth Joseph, Denver, Colorado (retired)
Rabbi, Hebrew Educational Alliance, Denver, Colorado (retired)

The spiritual odyssey of a liberal Jewish scientist who must keep asking, "Why?" A challenge to the believer and the nonbeliever alike.

STEVEN STEINBERG
Chaplain, Yale–New Haven Hospital
Rabbi, Quinnipiac University, New Haven

A careful and capable analysis of many religious ideas... for anyone who is willing to face reality [and] is no longer willing to delude themselves with magical religious solutions to real problems.

F. HENRY FIRSCHING
Professor Emeritus of Chemistry
Southern Illinois University–Edwardsville

For Deanna

David and Donna

Rachel

Alex and Noah

Table of Contents

To doubt everything or to believe everything are two equally convenient solutions; both dispense with the necessity of reflection.

JULES HENRI POINCARÉ

Preface

When I first circulated a draft of this book, a philosopher told me that my discussion of the Ontological Argument was so bad he could not help me improve it. Later, a physician read a slightly expanded manuscript and asked why I had wasted so much time on an obviously fatuous argument like the Ontological Argument. An engineer, reading the same manuscript, wrote in the margin, next to the Ontological Argument, "This is nonsense—are you sure you got it right?"

Similarly, a clergyperson told me that my scientific arguments were too detailed, a scientist told me that my religious arguments were too detailed, and another scientist begged me not to write any more about the Book of Ecclesiastes.

Thus, I confidently predict that the book will satisfy no one. I hope, rather, that it challenges all who read it.

My former student David Allen asked me precisely when I had started to study for this book. My immediate answer was, "I have always been studying for this book." In 1995, however, I decided to put my thoughts to paper.

As I note in the Introduction, my investigation brought me from science and philosophy of science to religion and philosophy, Biblical criticism, evolution and cosmology, mathematical physics, and the science of the brain. I do not have first-hand knowledge of many of these fields, so I have gone to the literature for my information. Perhaps that defines me as an amateur in the strictest sense of the word. I have read or reread over 100 books and countless articles in preparation for writing this volume. Some of these were written by scientists, some by clergypersons, some by philosophers and philosophers of science, some by journalists and science writers, and even one or two by novelists. Except for a handful of books and articles on physics and one statistics paper, every one is accessible to the diligent layperson; that is, anyone could read the same material as I read and draw his or her own conclusion. I present mine here.

I have tried not to rely on general knowledge but rather to document almost every statement of fact and to give citations wherever possible. I relied for some facts on encyclopedias, mostly the *Encyclopedia Britannica*; I do not cite facts taken from an encyclopedia. For historical facts, such as dates, I have relied also on *The Columbia History of the World*.

Several notes: All people in this book are real. When I use their real names, I identify them by both first and last names. When I use a first name only, that name is fictitious, but the person is not. To those who think they recognize themselves, I can only apologize and, as I did in *The Technical Writer's Handbook*, offer the hope that it is really someone else.

English unfortunately has no single word for *he or she*, *him or her*, or *his or hers*, so I have decided to stick with the traditional, masculine *he* when referring to God. I frankly thought that *she* would be a bit affected, and *it* would probably be offensive. The conception of God as male is a real problem, however, and I do not want to minimize it. Very simply, I am stuck with a limited palette of pronouns and have to do the best I can.

As for dates, I will not use A.D. and B.C., for "in the year of our lord" and "before Christ," because these imply a specific religious belief that not all readers share. Instead, I will use the scholarly but perhaps unfamiliar C.E. and B.C.E., for "Common Era" and "before Common Era." Thus, for example, the first Temple was destroyed in the year 586 B.C.E., and Constantinople fell to the Turks in 1453 C.E.

Finally, I will use the term *scriptural literalism* to describe any religious belief that is based on literal interpretation of a sacred text, such as the Bible or the Koran. It is synonymous with the more common term fundamentalism, but some consider fundamentalism a misnomer except when it is applied to certain Protestant sects. At any rate, scriptural literalism is a more descriptive term.

Acknowledgements

When you write a manuscript of around 100,000 words, you find out quickly who your friends are. Fortunately, I have a lot of friends, many of whom I have made while working on this book.

I especially want to thank Harold Kushner, author of *When Bad Things Happen to Good People* and other books, for his thoughtful and serious reply to my letter. David Allen, now with Corning, read an early manuscript with great care and made a host of valuable suggestions. Likewise, Edward Boosey, a wildlife conservationist in Norfolk, England, lavished 24 typed pages of comments on my manuscript. Bob Feuerstein, then Research Professor of Electrical Engineering at the University of Colorado, inspired the second-last chapter. Brendan McKay, Professor of Computer Sciences at the Australian National University, and Dave Thomas, Physicist at Los Alamos National Laboratory, contributed to the sections on the Bible codes and the equidistant letter sequences, and Jack Wang, Statistician with the National Institute of Standards and Technology, very kindly read the article about equidistant letter sequences and discussed it with me.

Mike Grant, Professor of Biology at the University of Colorado, explained some of the finer points of evolutionary theory very clearly. Warren Holmes, Professor of Psychology at the University of Michigan, likewise helped me with the question of incest avoidance. Tom Genoni, West Coast Bureau Chief of the Committee for the Scientific Investigation of Claims of the Paranormal, contributed to my discussion of near-death experiences. Michael Friedlander, Professor of Physics at Washington University in St. Louis, added considerably to my treatment of philosophy of science in Chapter 2. Ashley Johnson, then a student at Colorado College, helped enormously with the research. Alice Levine, a free-lance editor who lives in Boulder, contributed greatly to my organization of certain sections and offered invaluable advice whenever I needed it. Barbara McKinney of the Colorado School of Mines worked with me on the title.

Others who read and commented on the manuscript (not always favorably!) are Bradley Alpert, an applied mathematician at the National Institute of Standards and Technology; Leonard Boonin, Professor of Philosophy at the University of Colorado; John Carnes, Professor of Philosophy at the University of Colorado; Allen Cherin, a physicist at Bell Laboratories; Tim Drapela, Physicist at the National Institute of Standards and Technology; F. Henry Firsching, Emeritus Professor of Chemistry at Southern Illinois University; Elizabeth Fox, history teacher; Sidney Fox, a physicist at IBM; Don Jennings, retired Physicist at the National Institute of Standards and Technology; Marvin Gang, electrical engineer and stockbroker; Lloyd Gelman, physician; Daniel Goldberger, Traditional rabbi and pastoral counselor; Tony Leggett, Professor of Physics at the University of Illinois; Judah Levine, Physicist at the National Institute of Standards and Technology and sometime Talmudic scholar; Chris Mechels, a retired computer scientist from Cray Research and the Los Alamos National Laboratory; Steven Mechels, Physicist at TYCO Submarine Systems; Rena Segal, consultant; Paul Shankman, Professor of Anthropology at the University of Colorado; Steven Steinberg, Reform rabbi and hospital chaplain; David Wollman, Physicist at the National Institute of Standards and Technology; Jeffrey Yemin, computer scientist; Lowell Yemin, a nuclear engineer with Raytheon; Arthur Young, retired social worker; and Deanna Young, French teacher.

These worthy readers caught a number of errors and an even greater number of infelicitous uses of the language, and I am grateful to all of them. Whatever errors and infelicities remain, however, are mine and mine alone.

Boulder, Colorado, 2000

May it be your will, my God, that no misunderstanding or confusion arises through my teachings. May my colleagues and students value my learning; and may I value theirs.

NEHUNIA BEN KANA

Chapter 1
Sort of a hypothesis

I used to have a colleague I shall call Robin. He is a bright guy and a good scientist, and I think highly of him. He is also a member of a small Baptist sect and a Biblical literalist. Once, Robin owed me a favor, so I said, in essence, "Sit down. I would like to know why you hold your religious belief without evidence or, if you have evidence, what that evidence is."

We talked for the better part of an hour. Robin told anecdotes, talked about reports of "miracles" from all over the world, and spoke of his inner conviction, his inner feelings. I asked why he thought the religion of *his* parents was right and all others were (therefore) wrong. I asked if he would be a Koranic literalist if he had been born in Islamabad instead of Cleveland. He calls this my "accident of birth" argument, but he has no real answer to it.

Early on, I asked whether his belief was allegorical, that is, an approximation to the truth, or simply his way of getting at God and no better or worse than someone else's. Was his belief a hypothesis that he would employ as long as it worked, or was it absolutely true?

No, he answered, it is absolutely true.

At the end of the hour, he said, as best I can recall, "Look, what you said earlier, about being a hypothesis. [Pause.] I guess it is sort of a hypothesis." Saying so made him feel threatened. You could see it in his body language, hear it in his voice, see it in his eyes. So I quickly stopped the conversation.

1

The discussion with Robin reinforced my suspicion that relatively few people truly evaluate their religious beliefs. Most people I know have adopted the religion of their parents or one closely allied, if they have adopted any religion at all. Indeed, that must be generally so, for otherwise religions would be distributed more or less uniformly around the world. In fact they are not. Each religion is concentrated in its own area: Roman Catholicism and its descendants in western Europe and the Americas, the Orthodox religions in eastern Europe and northern Asia, and Islam in a band that reaches from Algeria to Indonesia, for example.

Thus, we may infer that most people simply adopt the religious practices of whatever culture they were born into; I have done that myself by adopting liberal Judaism.

I was raised in a secular Jewish household but attended Hebrew school at a Conservative synagogue on Long Island. I assumed that the Bible stories we were taught were just stories and did not believe any of them. That is not to say that they were without merit; myths and stories may have a great deal to teach us, even if they are not reporting historical fact. I merely thought that the Bible stories were not objectively true and continued to believe that for many years.

Although I attended synagogue occasionally in college, my connection with the Bible and religion lapsed until I joined a Jewish Reconstructionist seminar led by Leonard Rosenthal in Albany, New York. Later, when I was a professor of natural sciences at a small, now defunct college in upstate New York, Tom Davis, a Protestant minister and the chaplain at Skidmore College, helped me get through a couple of lectures on the New Testament. In the process, he convinced me that at least some Biblical stories were based on fact. Why, asked Davis, would the Bible present its heroes with serious flaws if the accounts were fiction? Why present Moses as a stutterer or King David as an adulterer if there were no truth to the stories? This kind of analysis was relatively new to me and sharpened my interest.

Since meeting Leonard Rosenthal and Tom Davis, I have become active in the synagogue and the Jewish community. I have had more than 25 years to read, attend seminars, and study the Bible and religion. In preparation for writing this volume, I read or reread more than 100 books and countless articles. I consider it a serious mistake to accept the Bible or any other dogma as absolute truth. After all, depending on how you count, there must be several thousand religions in the world. If only ten percent of them argue that they are the Only Right Ones, then all but one, or several hundred religions, must be wrong. Which ones?

A digression on terminology: When I write *belief*, I mean the belief in an objective *fact*. Certain beliefs may be held without evidence; these are *opinions*, such as my opinion that all people are entitled to a minimum standard of living. You may try to talk me out of holding this opinion, but you can find no crucial bit of evidence that will force me to change my mind, because an opinion is not a belief about a fact. By contrast, a fact is a statement that is necessarily true. If someone believes in a statement that is not true, as that the earth is hollow, that belief is plainly wrong. It is not always easy to decide what is a fact and what is not, but, at least in principle, you can compellingly refute a "factual" statement that is wrong. Belief in God is a belief in something that may or may not be a fact. Believing on faith alone ought not be satisfactory for people (whether scientists or not) who are used to relying on facts and interpretations of facts for all their other beliefs.

It is preferable seek evidence, to use the brain God gave us, if you will. One purpose of this book is to challenge conventional religious beliefs, to put them to the test, and to see how well they can stand up to the kind of scrutiny usually reserved for scientific hypotheses. It is aimed at those who are unimpressed by the current wave of mysticism and spiritualism and need to look for something else, at scriptural literalists who are willing to evaluate or defend their beliefs, and at those who think that belief in a supernatural is a prerequisite for religion.

3

Engineers often say, "A design or a computer program is never finished; it is just stopped for want of time." That is, you can never be certain that you have got a design or a program completely right, but at some point you have to make the decision that it is satisfactory and build the device or run the program. In the same way, we necessarily make all our decisions about everything on the basis of incomplete information; it is impossible to have all the facts. Facts are, nevertheless, important, and it is necessary to make an effort to get as many facts as possible before making a final decision, just as it is imperative not to stop a design or a computer program too soon.

In this book, I will examine the case for the existence of God. My examination will be based, as far as possible, on verifiable facts. It touches on science and philosophy of science, religion and philosophy, Biblical criticism, evolution and cosmology, mathematical physics, and the science of the brain. I grant that we do not have all the facts, but, like the engineers, we have enough facts to stop and draw our conclusion.

This book, then, has two main purposes: First to convince you, if necessary, that the only truly reliable knowledge is knowledge that is based on careful observation, reasoning, and experimentation. That is, beliefs are no more than hypotheses and need to be tested again and again, just like scientific theories. Sometimes, our beliefs will stand scrutiny; sometimes, they will not. But, apart from trivial statements like "That egg is white," no objective statement about the world should be accepted as fact unless it has been tested thoroughly.

My second objective, and the bulk of the book, will be to apply scientific method to the hypothesis that the universe is governed by a creator.

Science is empirical; that is, it is based on observation. In science, observation leads to hypothesis and hypothesis to testing. When hypotheses are uncontradicted after many tests, they are generally regarded as established facts. For precisely that reason, science has a greater claim to dependability and objectivity than other belief systems (Chapter 2). This statement is true because scientists do not assume that their theories are correct but rather test their theories in the real world. Science is self correcting: it not only recognizes its own mistakes, but actively looks for them and tries to correct them. That is, scientists draw conclusions from the available facts and modify their ideas when new facts are discovered. Scientists do not choose a hypothesis and assume that it is true; rather, they seek the best among competing hypotheses. Religion operates similarly in principle, I think, but in fact generally draws its conclusion first and then sets about to prove that conclusion rather than to examine it objectively.

Chapter 2 sets the stage. Chapter 3 asks, What happens if we apply scientific reasoning to belief in God or belief in miracles? People who believe in miracles use evidence selectively; they give God credit for good things but never assign blame for bad things. By ignoring the possibility that a "miracle" is no more than a coincidence, that is, by ignoring statistics, they ignore arguments that might refute their belief. Similarly, alleged signs from God are usually based on such devices as numerology or selective use of evidence. Indeed, I will show that an apparently fascinating statistical argument based on the Book of Genesis is based on such misuse of evidence. Finally, many postulate a God because, for example, they feel a need for a universal code of morality. Is this a valid argument or a matter of taste?

In Chapter 4, I ask whether the Bible is the work of God and therefore literally true in all details. Much of the Bible is poetry and is neither true nor untrue, whereas other parts seem to be recounting actual events. Some people classify the creation stories in the Book of Genesis as poetry or mythology, whereas others believe they are factually accurate. Careful analysis of the Bible, specifically the Book of Genesis, the Book of Jonah, and the Resurrection scenes in the Gospels, shows that the Bible cannot be accurate in all its details.

5

The existence of evil (that is, true evil, as opposed to random misfortune) is a major problem for those who believe in a God who is simultaneously benevolent, omniscient, and omnipotent. Chapter 5 shows that the Book of Job, though widely regarded as a profound study of evil, does not truly illuminate the problem of evil, but rather exemplifies our responses to misfortune. Our response to misfortune or evil is clear, whereas the origin of that evil is not. Chapter 5 explains what the author of the Book of Job could not have known: that much of what we call evil is the result of our biological nature.

To a philosopher, the "Arguments" for the existence of God are the Ontological Argument, the Arguments from First Cause, Contingency, and Design (Chapter 6), and the moral argument, which I cover in a different context in Chapter 5. The Arguments are more closely reasoned than the popular beliefs of Chapter 3 and the Biblical literalism of Chapter 4. Even the supporters of the Arguments agree that they have many problems, but some have seen evidence for intelligent design in the modern scientific theories of evolution and cosmology. In light of compelling evidence for periodic mass extinctions, the theory of evolution argues strongly against design, whereas cosmology provides evidence for a creation event but no evidence for a purposeful creator. Chapter 6 concludes by showing that religious or mystical experiences do not provide convincing evidence either for a creator or for a higher level of existence.

In Chapter 7, I draw my conclusion: The bulk of the available evidence does not support the hypothesis that the universe had a purposeful creator. Indeed, if the hypothesis is that the creator is also benevolent, then some of the evidence supports precisely the opposite conclusion. Biological systems were not created by God but are no more than complex physical and chemical systems. Descartes was mistaken in separating the mind from the body; what we call the mind is the result of natural processes that occur within the brain. I therefore doubt that we have free will in the strictest sense of the term.

In place of theism, Chapter 8 offers the knowledge that the universe is comprehensible, that we are responsible for ourselves, and that meaning and values come from within us. Why then am I religious in the broad sense that I observe many religious rituals and constantly study religion? Indeed, can you be religious without believing literally in God? That is, can you believe in God allegorically? Why do I attend synagogue, fast on Yom Kippur (the Day of Atonement), or host a Passover seder? I deal with these topics in the final chapter, which is a series of real conversations, as best I can reconstruct them. In these conversations, I attempt to answer questions I have been asked: Do you believe in God? Doesn't your scientific approach dehumanize us? Why do *you* pray, and, when you do, who listens? Why do I have to be moral? Where does faith fit into your worldview? Does it not create a void in your life to know that everything is in the long run pointless? Does not ethics follow from religion? My answers to these questions may not be entirely satisfying, but they are at least consistent with the best scientific knowledge of our time.

Where is the Tewa [an Indian nation] equivalent of television or the atom bomb? It is hard to imagine that given any amount of time the Tewa's way of explaining reality would have led to the digital computer or laser holography.

GEORGE JOHNSON

The problem is that most people never learn the difference between a good explanation and a bad one. Consequently they come to believe all sorts of weird things for no good reason.

THEODORE SCHICK JR.

The theory of evolution by natural selection is on its own sufficient to explain life.... By invoking the idea of evolution by natural selection as God's way of doing it, you are in effect invoking the one way which makes it look as if God wasn't there.

RICHARD DAWKINS

Chapter 2
Science, evidence, and nonsense

You will often hear that science and religion have nothing to say to each other—that they operate in different domains, that religion describes a realm about which science can have nothing to say. Nothing could be farther from the truth. As Einstein [1954] said, "Science without religion is lame; religion without science is blind." Religion addresses ethics, purpose, meaning. When religion makes certain factual claims, it is not necessarily impinging on the domain of science; nevertheless it is making statements that can be evaluated by the scientific method.

9

Religious beliefs that fly in the face of known fact are, frankly, wrong. Other beliefs, even if they cannot be evaluated with recourse to known facts, can still be examined, as opposed to simply accepted.

The Catholic church is not alone when it says that faith is superior to reason. How many times have we heard someone say you just have to have faith! Why, I wonder, do they want to have faith in something that may well not be true? Why not regard their beliefs instead as hypotheses to be tested? Why assume the answer to what is in some ways the most important philosophical question of all time?

Why indeed? In this book, I will test some of the hypotheses of religion. I will implicitly examine the claim that there are "other ways of knowing." I will try to show, in particular, that the scientific method—careful observation, experimentation, and reasoning—or something like it is the only reliable way to assess claims of fact and that failure to apply scientific reasoning often leads to error and misunderstanding. Indeed, seeking evidence to support a religious belief is not at all odd: Those who rely on testimonials or anecdotes to reinforce their faith tacitly agree that evidence is important; what is at issue is the quality of the evidence, not the need for evidence or the relevance of evidence.

Scientific method

Let us begin with a statement that some will consider so obvious as to be fatuous:

There is objective reality.

By this, I mean not that objects outside ourselves exist but rather that truth can be determined by careful experiment or observation and does not depend on our opinion of it nor, indeed, on our understanding of it. I raise this point because there is a widespread misunderstanding that science is just another belief system, whose truths are no better or more reliable than those of

10

any other belief system. That is, since all knowledge is imperfect and perfect objectivity is impossible, scientific claims are no more dependable than any other claims. People who believe this statement seem not to understand that there are degrees of dependability or objectivity. As the science writer Timothy Ferris [1995] has stated it, scientists do not ask us to take their words for what is real. Instead, they appeal to experiment and observation. Scientists are not always right, but people who have studied a subject empirically have a better chance of being close to the truth than those who have not. The more highly developed sciences, such as physics, chemistry, and molecular biology, can make certain claims with near certainty. These claims are properly called "scientific facts."

Why then do some people reject what are almost certainly scientific facts? The etiology of this disease is complex. No doubt, it arises in part from a backlash against some of the misuses of science (both advertent and inadvertent) by some of its practitioners. But, more likely, it descends from our recent penchant for historical and cultural revisionism. Revisionist historians, for example, argue that history is written by the victors, whether of wars or elections. The victors depict themselves as the good guys and their enemies as the bad guys. A generation or more later, revisionists may challenge the conventional interpretation of events. They are not always right, but neither are they always wrong. Sometimes, however, the revisionist doctrine takes hold and it almost seems as if we have turned history on its head; contrast, for example, the present belief that Columbus was a scoundrel with past reverence for Columbus as a brilliant innovator. This is often called a "paradigm shift," even though this use of "paradigm" was invented by Thomas Kuhn to describe scientific revolutions (see "Paradigm shifts," this chapter). The idea of a paradigm shift has left many believing that scientific beliefs are not governed by evidence but rather by the culture of science, in the same way that political, historical, or economic beliefs sometimes seem to

11

depend on political point of view or ethnic or religious background.

In addition, many mistrust science for purely political reasons. For example, the well known social critic Barbara Ehrenreich and her colleague Janet McIntosh tell an anecdote about a social psychologist who presented a paper to a seminar on emotions. [Ehrenreich and McIntosh, 1997] As soon as the psychologist used the term "experiment," she was derided for the fact that experimental method was developed by white Victorian males. She admitted that white Victorian males had wrought a lot of damage but noted that experimental method has also had a lot of success, such as the discovery of DNA. Someone from the audience jeered, "You believe in DNA?"

In short, certain critics of science perceive science as a white man's culture, where the stress may be placed on *white* or *man*, as you please. In their eyes, it is therefore not only to be mistrusted but also completely erroneous. This is an example of what is sometimes called the *genetic fallacy*: evaluating (in this case rejecting) a statement or an argument because of its origin, or genesis, and subsequent development, and not because of what that statement or argument suggests now. The genetic fallacy is commonly applied to religious belief, as in the statement that religion had its origin in primitive peoples' fears of the unknown. Even if that statement were true, it would shed no light on the truth or falsity of religious belief. Similarly, even if science were a white man's culture, the fact would shed no light whatsoever on its truth or falsity.

Finally, the modern (or postmodern) literary theory of deconstruction has contributed to the view of science as an arbitrary belief system. [Gross and Levitt, 1994] Deconstruction treats all knowledge as a series of "texts." As Dan Carter, [1992] Professor of History at Emory University, phrased it, these texts are considered to be "divorced from reality." Their content is assumed to depend entirely on the prejudices and preconceptions of their authors. Evidence is therefore assumed to be irrelevant to the conclusions drawn in any text whatsoever. Truth is relative. In the words of Lewis Vaughan, [1997] an editor of *Free Inquiry*, if you sincerely believe that the earth is 6000 years old, then, for you, the earth *is* 6000 years old, and there is no need to demonstrate the truth of your belief. The problem with this reasoning is that it makes each of us infallible, even though we disagree with each other.

Students and critics of fiction often treat fictional characters as if they were real people and go about "psychoanalyzing" them, or trying to figure out what motivated them. You can reasonably interpret fiction in ways that differ from the author's original intention; you may also be able to ignore that author's culture and background, though that is more problematic. You can to a lesser extent do the same thing in history or sociology. But you cannot interpret any text any arbitrary way. As my friend Debbie Taylor, a high school English teacher, says, indignantly, "If you think Little Red Riding Hood wore a blue cape, *you're wrong!*" Likewise, if you think Goldilocks was a brunette, you're wrong. If you think that Cyrano didn't write the letters to Roxane, you're wrong. You may argue, if you want to, that he didn't really love her, but you cannot sensibly argue that he didn't write the letters. Within the context of the play, that is a fact, and any analysis of the play has to be at least consistent with the facts of the play. Similarly, attempts to interpret a text from a different era independently of the outlook and values of its author will often lead to anachronism: Hester Prynne, in Hawthorne's novel *The Scarlet Letter*, has a child out of wedlock; was she (or Hawthorne) therefore an early feminist? The idea is silly. Even dramas or works of fiction cannot be analyzed wholly without regard for their origins or their internal logic.

Similarly, it is fatuous to treat the physical sciences as texts divorced from reality or reject scientific discoveries as constructs that depend on the core beliefs of the scientists. "Laws," such as the law of gravity, are firmly grounded and based on observed fact, not dissociated texts. All you can hope to do is to find a better law, that is, a law that more closely describes experimental

13

observation or has a wider range of validity than the old law. Even if you do so, however, your new law must be exactly equivalent to the old law in those areas where the old law is known to be accurate. By "exactly equivalent," I mean that the new law must predict results that are immeasureably different from those predicted by the old law. Relativity, for example, yields results that are equivalent to Newtonian mechanics when a particle's velocity is low, and quantum mechanics yields results that are equivalent to Newtonian mechanics when a particle's mass is large. If they did not do so, then they would be recognized as wrong, and no amount of deconstruction would make them right.

In short, scientists believe what they do because it has been shown to be true, or at least true within certain limits, not because they are a community of shared values such as a religious or political community. That scientists are a community of shared values is arguably true, and their shared values perhaps determine what they study. It does not, however, determine the truth of what they discover. Indeed, if you truly think that the beliefs of scientists are as open to interpretation as political beliefs, then I suggest that you test-fly an airplane designed by a postmodernist who did not believe in Bernoulli's principle or have your next major surgery performed by an English professor who does not believe in the germ theory of disease.

Sometimes scientists themselves contribute to public confusion and make it look as if we can neither agree nor make up our minds. This is more often true in, for example, medicine and certain social sciences, for the simple reason that the number of variables is so great and uncontrollable that those endeavors are very much harder and less exact than physics and chemistry. Journalists add to the confusion when they report preliminary results incompletely and without qualification, as when they contrast the results of a small study with those of a much larger study, and fail to report follow-up studies.

14

Consider the recent confusion about unsaturated fats. [*UC Berkeley Wellness Letter*, 1996] For a good many years, we were told to eat hydrogenated vegetable oils (such as those in margarine) because they were supposedly better for us than animal fats (such as butter). Then it appeared that hydrogenated vegetable oil was not so healthful after all and that, maybe, butter was preferable to margarine. Press reports might leave you with the uncomfortable feeling that scientists have changed their minds and are now stating their new conclusions with as much certainty as they stated their old conclusions.

This sort of thing happens, in the words of Neil deGrasse Tyson, [1998] the Director of the Hayden Planetarium, "because the frontier of discovery is a messy place." Scientists typically publish a tentative conclusion based on preliminary data. Slowly, the data are refined, by the original scientist and others, and finally a much firmer conclusion may be drawn. Years often pass between the tentative conclusion and the firm conclusion. The firm conclusion may confirm the tentative conclusion, or it may partly or wholly disconfirm it. Yet the press often reports a tentative conclusion as if it were a firm conclusion, even though the scientist did not present it as a firm conclusion. In this way, the press sometimes leads people to believe that science is not a rational or trustworthy process.

In the case of the saturated fats, scientists have never once wavered from their advice to eat far less fat and oil, and to restrict the oils we eat to vegetable oils. Scientists still believe that unsaturated fats are more healthful than saturated fats, but there is evidence that certain saturated fats are more harmful than others. Only the ranking of the fats has changed. In short, scientists act on the total information that is available to them; as that information changes, scientists refine their advice.

* * *

Some people seem to believe what they believe simply because they believe it. A woman I shall call Ricky once told

15

her uncle Arnold that the Spanish Inquisition had burned 15 million women at the stake. Arnold considered that a surprisingly large number and asked her for verification. According to Arnold's version, Ricky said, "If I believe it, then it is true." Arnold replied, "Then the world was once flat, but now it is a sphere." Ricky was so distressed by having her belief challenged that she got up and left the room.

Ricky has fallen into the trap of thinking that anything she hears is true as long as it is congenial to her to believe it. That is, if she *feels* that something is true, if it resonates with her, then she believes it without asking for proof. She is by no means alone; this is the manner in which many people determine their beliefs. (The *Encyclopedia Britannica*, by the way, estimates that 2000 people were burned during the 15 years that Tomás de Torquemada was the Grand Inquisitor in Spain. Even if 10,000 were burned in all of Europe during that same 15 years and the rate of burnings remained constant, approximately 150,000 would have been burned in 200 years; the rate would have had to increase by a hundredfold for 15 million to have been burned in 200 years. The author and entertainer Steve Allen [1990] cites sources that estimate 300,000 women burned by both Catholics and Protestants as witches between 1484 and 1782. Later, Allen cites estimates in the millions, so Ricky may not be as far off as Arnold thought. The majority of those burned as witches, incidentally, were women.)

Those who believe in astrology, clairvoyance, past lives, UFO abductions, mental telepathy, foot reflexology, and the like are not the only ones who put their trust in faith rather than reason. Here is an example from medicine.

Homeopaths are followers of Samuel Hahnemann (1755-1843), a German physician who, as *Consumer Reports* [1994] puts it, practiced at a time when medicine used "drastic" methods like "bloodletting, blistering, and purging," typically with highly toxic mercury compounds. Hahnemann instead tried herbs, minerals, and other substances according to a "law of similars" that dates back at least to the ancient Egyptians. This "law" states that "like cures like" and led the Egyptians, for example, to eat ostrich dung (or possibly a plant called ostrich dung) in an attempt to cure diarrhea. Hahnemann was slightly more sophisticated: He believed that certain substances could cure symptoms precisely like those they caused. The "proof" is by analogy: Quinine causes headaches, thirst, and fever; these are the symptoms of malaria; quinine cures malaria; QED.

Unfortunately, some of the substances that Hahnemann tested were very toxic, so he diluted his mixtures more and more but still thought he saw a curative effect. Ultimately his mixtures became so diluted that almost certainly not a single molecule of the "medicine" remained in the "solution," which nevertheless contained many parts per million of impurities.

Hahnemann lived at almost exactly the same time as John Dalton, the founder of the modern atomic theory of matter, and apparently he was well aware that matter was not a continuous entity that could be diluted forever. That is, he knew that his "solutions" contained literally none of the "medicine." [Garnett, 1995] Instead, he believed that the solution had to be shaken vigorously each time it was diluted in order to "potentize" it and leave in the water the "spirit-like" essence of the medicine. In effect, Hahnemann said, the water remembered what had been put into it.

Hahnemann lived at a time when medicine was in its infancy. He understood neither the placebo effect nor double-blind tests, where neither the physician nor the patient knows whether the patient is receiving a medication or a placebo. That is, he did not realize that patients can sometimes appear to respond to a worthless drug or even a sugar pill, just because they believe it to be efficacious. This effect is so strong that sometimes 30 percent of those who receive a placebo can appear to be cured by it. In short, it is very easy for either a patient or a practitioner to be deceived by a "cure" effected either by a disease that has run its course or by the placebo effect. Still, considering how much Hahnemann was diluting his solutions, I

17

find it hard not to argue that he should have realized early that his treatments were just not effective.

Even though we still have practitioners of homeopathic medicine, I think of them as alchemists. The alchemists spent much of their time trying to turn base metals, such as lead, into gold. Today, we know that that is impossible; you can make chemical compounds, but you cannot turn one element into another by means of a chemical reaction. Nevertheless, the alchemists were the forerunners of modern chemistry, rather as the astrologers were the forerunners of modern astronomy. Is Hahnemann's work relegated to history? Do we think of Hahnemann as we think of the alchemists or the astrologers? Alas, no.

Homeopathy is considered medicine in the United States and is practiced by medical doctors and osteopaths. What do its practitioners say about the dilution problem? In effect, that the water remembers. One of the leading practitioners, Beverly Rubik of Johns Hopkins University, has been quoted in several places as saying, "We need to rethink the properties of matter. It's that deep." Rubik, who proposes no new theory of matter, does not understand that a new theory has to be consistent with current theory. That is, it is extremely difficult to overthrow a theory that is known to work; usually, the best you can do is to extend such a theory. Homeopathy, as understood by Rubik, does not extend modern chemical knowledge; to the contrary, it denies our present understanding of the chemical bond.

Does the water remember? Jacques Benveniste, a researcher at the University of Paris, thinks so. Benveniste mixed a certain antibody in water and exposed white blood cells to the solution. The antibody causes the cells to release the compound histamine, which is stored in granules in the cells. Cells that have released their histamine are *degranulated* and absorb a certain red dye differently from cells that retain their granules. The difference can be seen under a microscope. [Friedlander, 1995; Lawren, 1992. Michael Friedlander is Professor of Physics at Washington University in St. Louis, and I will have several

18

occasions to refer to his book *At the Fringes of Science*, a study of scientific error and fraud, and pseudoscience.] According to Benveniste, cells were degranulated even when the solution was so highly diluted that it contained no antibody whatsoever. The water had remembered. Additionally, the degree of degranulation oscillated with every tenfold dilution of the solution; this claim made Benveniste's story especially hard for many scientists to take seriously.

Despite critical referees' reports, Benveniste's paper was published in *Nature*, one of the premier scientific journals in the world. A furor ensued, and *Nature* dispatched its editor, John Maddox; the magician James Randi; and Walter Stewart of the U.S. National Institutes of Health to investigate. Randi may seem like an odd choice, but he has a reputation for exposing pseudoscientists and, as a magician, is an expert on deception, whether it is deliberate or not. Stewart had been one of the referees of the original paper and had earlier been involved in a controversial search for error or fraud in science.

The team concluded that Benveniste's results were a consequence of inadequate controls. Specifically, they concluded that the people who evaluated the microscope slides knew what they were expected to find—and found it. This is sometimes called *experimenter bias*: The observer or experimenter knows what to look for and seems to find it, whereas an observer who is untainted by such knowledge does not. When Maddox's team arranged a blind test, the results were negative. Benveniste's results look like a classic case of self-delusion.

According to Lawren's sympathetic article, however, Benveniste is continuing his research with better controls and has gotten similar results. He has a theory that molecules communicate by electromagnetic radiation instead of by exchanging chemicals. The radiation remains in the water, even after the molecules are removed by dilution. His theory is an arm-waving theory; that is, it is not quantitative. It does not explain why the radiation is not diluted as the solution is diluted

19

nor what is the mechanism that prevents the radiation from escaping from the solution. Neither does he have evidence for the existence of such radiation, which would have to be greatly different from any known electromagnetic radiation if it were to remain confined in the solution, rather than to escape at the speed of light. Finally, his theory makes no predictions that might allow us to test it.

* * *

You can distinguish a pseudoscience such as homeopathy from true science in several ways. Sometimes, the pseudoscientist proposes a bizarre theory that is contrary to known fact, as in the statement that the water remembers. That is, the pseudoscientist has a thesis that does not build on old, known facts; rather, everyone else is out of step and has to rethink the old, "wrong" theories. Rather than apply well-established principles to their own beliefs, Rubik and Benveniste think that it is the molecular chemists who have to reexamine well-established principles in light of weak or nonexistent evidence from homeopathy.

I am not arguing here that unexpected discoveries are impossible, nor that the conventional wisdom of scientists is always right. The discovery that stomach ulcers are most often caused by a bacterium flew in the face of conventional medical wisdom, for example. [*Harvard Health Letter*, 1994] But such discoveries, while surprising, do not require a radical overhaul of science at the most fundamental level. Bacterial infections were not unknown at the time, and the evidence that ulcers were caused by emotional stress was never very strong. No one was asked to reevaluate either molecular biology or psychiatry just because ulcers turned out to be the result of a bacterial infection. In addition, it remains possible that some ulcers are caused or exacerbated by stress, as physicians had originally thought.

Another distinguishing characteristic of pseudoscience is an observed effect that is so slight that it can barely be detected.

20

When it is not detected, the pseudoscientist offers a flimsy excuse, which we will later identify as an *ad hoc hypothesis*. Benveniste's electromagnetic radiation is an *ad hoc* hypothesis (see "Falsifiability," this chapter).

Finally, real scientists present evidence intended to support their theories. As Michael Friedlander pointed out to me, pseudoscientists invert the obligation. That is, they present an unfounded thesis and defy you to disprove it. They have the cart before the horse: the obligation of proof is on the people who make the claims. You have to demonstrate at least that your claims are plausible, for example, to get your ideas published in a respectable scientific journal. But pseudoscientists often deny the validity of accepted knowledge and then defy you to prove them wrong.

This is the tactic taken by Michael Drosnin, whom we will meet in connection with the Bible codes (see "Signs," Chapter 3). Drosnin's argument is no more and no less than "You can't prove me wrong." My relative Jack used to make the same argument when his belief in herbal remedies was challenged. Neither Jack nor Drosnin understands that you cannot ever disprove certain kinds of arguments. (These are claims such as R. M. Hare's *bliks*; see "Is religion falsifiable?" in this chapter.) Usually they are negative statements like Drosnin's or Jack's. I can never prove that an infusion of some herb did not cure Jack's cold; all I can do is cite evidence that the same herb worked no better than a placebo in careful studies. In the same way, I can never prove that Drosnin's theory is wrong; all I can do is show that the patterns he discerns in the Bible could have been caused by chance—appear no more often than is predicted by chance—and that his interpretation is not the only possibility.

* * *

When I was in college, I read some works of the parapsychologist J. B. Rhine. I told some friends that I had recently seen an engineer using a divining rod to douse for water.

21

I was skeptical, but he had located (or seemed to locate) an old dry well. A dry well is a large hole, often filled with rocks, into which water is funneled, stored, and allowed to seep away. For example, the downspouts from your roof may channel excess water from the roof into a dry well in order to reduce the quantity of surface water during a rainstorm. The dry well for which the engineer was searching had been disconnected from the downspouts, and no one knew where it was.

At any rate, my friends proposed a test. They would fill several similar bottles with water and leave some empty. All the bottles would be wrapped in identical towels and distributed in no special order over two beds. I would douse for the bottles that contained the water and find them. At first, I did!

That is, I did until Gordie walked into the room. He laughed, he sneered, he ridiculed. I could no longer find the bottles that contained the water. What did we say? No more than half believing it, we claimed that Gordie was somehow interfering with the experiment, that his disbelief was interfering with my ability to locate the water. After all, if you believe that dousing is some form of mental telepathy, you must also believe that someone else's brain could interfere with your own. We kicked Gordie out. But after that, I still could not find the water, and we concluded that the initial success had been a fluke. Note, though, our first reaction to Gordie's criticism: We attempted to patch up our theory with an *ad hoc* hypothesis.

What about the engineer? The dry well was empty, since it was no longer connected to its only source of water, which was the downspout. We knew where the downspout had been and therefore where the dry well probably was. The engineer probably found the dry well because he really knew where it was all along, though he certainly thought that he had found it by dousing. He did not find it all odd that he had doused for water and found an empty dry well.

* * *

In contrast to pseudoscience, real scientific theories build on old theories in the sense that relativity and quantum mechanics give precisely the same results as classical physics under certain conditions; in physics jargon, relativity *reduces to* classical physics in the limit of small velocity, and quantum mechanics, in the limit of large mass. Thus, even revolutions do not always overthrow old theories; most often, they encompass them or incorporate them. That is, even scientific revolutions are in some sense merely advances that extend and build on the knowledge of the past. Homeopathy, by contrast, is completely inconsistent with any known physics or chemistry.

Here is an example that stands in stark contrast to homeopathy and to my excursion into dousing. The nutritionist Phyllis Crapo of the University of Colorado Health Sciences Center once attended a lecture on diabetes. [Kolata, 1983] Diabetes is a condition of high blood sugar, and diabetics were then advised to avoid eating sugars but not starches. The (unnamed) lecturer, however, suggested that eating starches might be just as bad for diabetics as eating sugar, because our bodies have a great number of enzymes that quickly break down starches into sugar. Crapo was skeptical. For years, she had been advising her diabetic patients to eat starches and stay away from sugars.

Rather than dismissing what she had heard, however, Crapo set out to test her hypothesis. She and her colleagues fed foods that contained different starches and sugars to volunteers, some diabetic and some not. Then they monitored the concentration of sugar (glucose) in the volunteers' blood. To Crapo's surprise, she found that some starches were metabolized so fast that eating them was equivalent to eating sugar; that is, they caused a fast and dangerous rise of blood sugar. Potato starch was the worst in that regard and was almost as bad for a diabetic as eating pure sugar. Ice cream, by contrast, did not cause a rapid rise of blood sugar, probably because the fat in the ice cream inhibited the assimilation of the sugar. The findings of Crapo and others have since been used to modify the advice given to diabetics.

23

What is important in this story is that Crapo did not set out to *prove* her own hypothesis nor to disprove the speaker's. Rather, she set out to *find out* whether the speaker was right, not to *disprove* him (or by the same token to prove herself right). That is, unlike Hahnemann and his followers, she took an open mind into the experiment and tried to find out what is true. Those who try *to prove that something is true* will often succeed, even if the evidence does not support them. How? They will explain away results that do not support their thesis. When Maddox's team forced Benveniste to eliminate experimenter bias, Benveniste was not surprised at the null result; many times, he said, he had gotten a null result. [Lawren, 1992] Apparently, the water sometimes forgets, after all. We will discuss this ploy under the heading "Falsifiability."

Internal reality

I have read that believing in a reality external to yourself is a bit of a leap. Believing in minds other than yours is an extrapolation. True, such statements drive home the fact that everything we think we know is an inference. Still, I do not find this sort of philosophy particularly helpful. If you really do not believe in external reality, then (taking my cue from a famous quip by Samuel Johnson) I invite you to go to the nearest brick wall, smash your fist into it as hard as you can, and observe the results. If your fist is undamaged, then I will reconsider my belief in external reality in the light of the new evidence.

As for minds other than my own, I do not think that I am nearly clever or talented enough to have composed the symphonies of Mozart myself, nor written the sonnets of Shakespeare, nor painted the Mona Lisa. These were created by remarkably talented people and are not figments of my imagination, as I would have to assume if I rejected the existence of external reality.

Nevertheless, I argue that

There is also internal reality.

What I mean by internal reality is what you see and hear, think and feel. Internal reality includes your conception of the external world and your emotions or feelings. If you honestly say, "I love you" or "I like that painting" or "I have a headache," then it is true almost by definition. That is, your feelings are yours, and no objective test can deny them. We cannot apply scientific method to statements about internal reality. If you claim to have a headache, we can perhaps examine your behavior and seek clues to suggest that you are lying, but we cannot conclusively prove that you do not have a headache. Your statement that you have a headache is not a statement about the external world and makes no claims that anyone else can truly verify.

We can, however, apply scientific method to external reality. If you say, "The moon is made of green cheese," then it is not true, no matter how firmly you believe it. Your perceptions are a guide to external reality, but they are not foolproof.

* * *

I belong to a group called the Rocky Mountain Skeptics. Some of our members were involved with debunking a medical "treatment" called Therapeutic Touch, which was (and is) being taught at the University of Colorado Health Sciences Center. Let it suffice to say that Therapeutic Touch, which involves no actual touch, is said to direct energy fields from the practitioner to the patient. There is no objective evidence for the existence of such energy fields and no clinical evidence that Therapeutic Touch is more effective than a placebo.

At any rate, Rocky Mountain Skeptics became either famous or notorious, depending on your point of view, for petitioning the Regents to do away with the courses in Therapeutic Touch. At about the same time, I was asked to sit at the Skeptics' booth at an outdoor festival. Someone came and shouted at me, with no audible punctuation in his voice, and as best we could reconstruct it,

> I don't have to be a skeptic because I believe what I experience and no one can tell me what to think about it because I know what I experience.

In other words, if he believes something, then it is true. He then asked me if I had ever undergone Therapeutic Touch (no) and, as someone put it later, virtually rejected all knowledge, such as book-learning, that does not come from direct experience. Indeed, he seemed to think that only his perceptions were real and that no hunch or feeling needed verification beyond the evidence of his senses. If Therapeutic Touch seemed to be effective, then it *was* effective, and a person who has never experienced it has no right to criticize it. By this person's logic, I had to have been at Valley Forge with Washington in order to have the right to an opinion about what happened there.

If you agree that your perceptions are necessarily accurate indications of reality, then look at Figure 1, upper. This is known as the Müller-Lyer illusion. [Block and Yuker, 1989; Wolf, 1996] As far as I know, everyone perceives line segment (a) to be shorter than line segment (b). Indeed, as Block and Yuker point out, the illusion is so strong that it persists even when you add a ruler to provide incontrovertible evidence that the two lines have the same length, as is shown in the Figure 1, lower.

Can you trust the evidence of your senses? Of course you can, but only to a point. After all, you can easily be fooled, and some things are undetectable with your unaided senses. Sometimes, therefore, it is necessary to perform a measurement or a scientific experiment to be sure. Those who, like Hahnemann the practitioners of Therapeutic Touch, rely on their perceptions

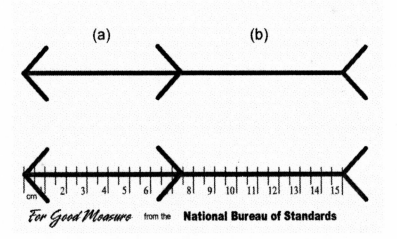

Figure 1. The Müller-Lyer illusion. The two line segments (a) and (b) are the same length, but (b) appears longer than (a). The illusion persists even when the scale shows clearly that the two line segments are equal in length. Your perceptions are not necessarily an accurate guide to reality.

alone are in too great a danger of being proved wrong when their hypotheses are tested rigorously. I do not consider it acceptable to argue that patients' experiences are real and may be valuable nevertheless. The experience of being fooled by a prestidigitator is real (and in some sense valuable), but the prestidigitator did not do what you thought he did when you watched him. Fortunately, a belief in a prestidigitator is harmless, whereas substituting homeopathy in place of competent medical treatment can be lethal.

Anecdotal evidence

By anecdotal evidence, I mean casual observations, stories, or testimonies. It is as easy to be fooled by anecdotal evidence as by an optical illusion. Consider, for example, the widespread belief that sugar causes hyperactivity in children. My friend Morris, a trained psychologist, believes it. His evidence is that children act wild at parties and on Hallowe'en.

Let us assume, first, that there really is a correlation between sugar and hyperactivity. Many people do not understand that mere correlation does not prove causality. Those who automatically assume that marijuana smoking causes heroin addiction, for example, should note that most marijuana smokers drank milk when they were children; I would wager that the correlation is nearly 100 percent. Yet no one would argue that milk drinking causes marijuana smoking. Similarly, it is necessary to prove whether or not marijuana smoking leads to heroin addiction, not just to assume it because of a possible correlation. The proof would consist of careful statistics leading either to a correlation or lack of correlation. If there is a correlation, it is still necessary to establish causation; it is always possible that whatever causes people to smoke marijuana also causes them to take heroin, but that the marijuana smoking itself does not cause heroin use. For example, it is possible that the underlying cause of both marijuana and heroin smoking is some other factor such as rebelliousness or a propensity for risk-taking. (Although some call marijuana a "gateway" drug, the evidence is equivocal. Most heroin users have used marijuana, but most marijuana users never use heroin. Similarly, tobacco smokers are far more likely than nonsmokers to use alcohol or marijuana. In neither case is there a clear causal relationship, however. [*CQ Researcher*, 1995; Ravage, 1994; Brooks, 1993])

Even if correlation necessarily proved causation, we would not always know which was the cause and which the effect. More important, though, it is possible that the correlation is incidental and that the cause was a third variable. Lightning

28

(that is, the visible flash) precedes thunder, but it does not cause thunder. Rather, an electric discharge causes both lightning and thunder. The light and the sound are incidental. To borrow another example, suppose that certain storks always return to Sweden in the spring. [Taylor, et al., 1995] Calves, lambs, and other animals are born in the spring. You might conclude, therefore, that the storks bring babies. Closer examination reveals, however, that babies are created by another mechanism and that the two phenomena are both separately related to the arrival of spring. That is, the *covariance* between storks and babies is caused by another variable, and, in fact, storks and babies (except for baby storks) are not causally related to one another.

Something similar seems to be the case with sugar and hyperactivity. According to Tufts University's *Diet and Nutrition Newsletter* [April, 1994], an experiment on sugar and hyperactivity has been done, and sugar is very probably not linked to behavior, at least not strongly enough to be detected easily. Briefly, researchers chose two groups of children in several age groups. Some of these children had been classified by their parents and teachers as "normal" and the others as "sugar-sensitive." The researchers went as far as to remove all the food from the children's homes and bring in food for three weeks. Some of the children received food with sugar; some, with the artificial sweetener aspartame; and some, with the artificial sweetener saccharin. Until the end of the experiment, no one knew which children got which food: not the researchers, not the teachers, not the parents, not the children.

Using over 30 different criteria, the parents, the teachers, and the researchers evaluated the children's behavior for the duration of the experiment. Except that some of the children who turned out to be on the sugar diets tested a little *better* on learning tasks, the research found no significant differences among any of the groups. One such study is not conclusive, but it certainly suggests that the effect of sugar on behavior is slight at most.

Why then do so many people believe that sugar affects behavior? My first thought was that the children often get their sugar in high doses at parties or during recess: precisely those times when they might be expected to act wild anyway. Possibly, the wild behavior is related to the environment, and the sugar is only incidental, just as the arrival of the storks is incidental to the birth of babies. There is a correlation between the sugar and the behavior all right, but the cause of the behavior is not the sugar but rather the party or the recess. The Tufts Newsletter suggests, to the contrary, that the very belief of the parents may cause a self-fulfilling prophecy: "Eat that candy bar and you'll be bouncing off the walls," as Tufts put it. The children naturally oblige by bouncing off the walls.

I cannot strongly enough stress the need for evidence. A neurologist at Tufts points out that it can be downright dangerous to blame sugar if a child really has a disorder, and the disorder goes undiagnosed because of the incorrect association with sugar. This is one of the reasons I find it very troubling that some people will believe outlandish claims about drugs or diets and on their basis forgo medical help. Indeed, sometimes the alternative "treatment" can prove dangerous or fatal. [Carter, 1996]

Is anecdotal evidence always wrong? By no means. For all I know, it may be right more often than not. But it should never be regarded as conclusive. Rather, anecdotal evidence should give you a hunch or allow you to formulate a hypothesis, *which you later test*, either by performing an experiment or, in cases where facts are readily available, by looking it up. The experiment, by the way, has to be carefully controlled and must not rely on your more or less casual observations. It is not true, as some wag has said, that the plural of anecdote is data. To the contrary, the plural of anecdote is incomplete or biased data, because we often ignore anecdotes that do not conform to our expectations. Scientists may have preconceptions that allow them to be fooled as easily as anyone else; they must take great pains to ensure that their experiments are designed so that their preconceptions cannot interfere with their conclusions. Regarding anecdotes as anything more than hints is a recipe for self-deception.

Does Morris now agree that the evidence at least suggests that sugar is not the culprit? No. Morris may be a trained psychologist, but he is almost impervious to evidence; he believes what he believes on faith alone. Instead of refuting the evidence, he introduced the factoid that sugar causes hypoglycemia and therefore is not without physiological effects.

Falsifiability

A theory must be *testable*, at least in principle, or it has no value. That is, a theory must suggest an experiment, an observation, or a measurement that will refute that theory if the experiment gives the "wrong" result. (By "wrong" result, I mean wrong from the point of view of the theory; a more accurate term might be "unexpected." A good experiment never fails; rather, it gives a result you did not expect. In scientific jargon, however, that is a wrong result. You might want to speculate whether such jargon reinforces our preconceptions.) A theory that is constructed in such a way that it cannot be refuted also cannot be tested and is therefore not a valid scientific theory.

For example, Einstein's General Theory of Relativity did not predict merely that star light would be bent if it passed close to the sun. More precisely, it predicted that it would be bent by a specific angle, and that angle was about twice what a naïve estimate based on a nonrelativistic particle theory of light would have predicted. Thus, when Arthur Eddington mounted his famous expedition to view a total eclipse of the sun, he had to *measure* the deflection of the star light, not just *note* it. If he had got the "wrong" answer, Einstein's theory would have been refuted. His getting the "right" answer, however, did not prove Einstein's theory, but merely provided supporting evidence in its favor. This is a crucial distinction: An overwhelming amount of evidence can never prove a theory. However, a single "wrong"

31

result can disprove it or, at least, cast it into serious doubt. That is, to be useful, a theory must be subject to refutation. In scientific terminology, this is called *falsifiability*. All good theories are falsifiable. This means that they can in principle be disproved, not that they are wrong or fraudulent. A theory that could not be disproved, even in principle, is necessarily too vague to have value.

Sir Karl Popper, [1968] Professor of Philosophy at the University of Canterbury in New Zealand and London University in England, was one of the most influential philosophers of science of the twentieth century and was knighted for his contributions. Popper argues that scientists do not record a lot of data and then recognize the right theory by induction. Indeed, he denies the idea that truth is in any sense obvious. Rather, he says, a scientist makes a conjecture and then tests that conjecture by subjecting it to rigorous experimental tests. Newton, for example, did not infer the Law of Gravity from Kepler's observations. Rather, he guessed at the law—that the attractive force between two bodies is inversely proportional to the square of the distance between them—and then showed that his guess led directly to Kepler's laws. Like a lot of philosophers since his time, Newton was troubled by the fact that his law involved *action at a distance*, that is, that two bodies are attracted to each other even though there is no obvious mechanism for that attraction. Nevertheless, the Law of Gravity is extremely accurate, and Newtonian physics is used to this day for almost all astronomical calculations. A deeper understanding of gravity had to await the development of Relativity, which accounted for the action at a distance that so troubled Newton.

Popper further argues that a good theory must be very precise: It must make specific, even daring predictions, not vague general statements. For example, you could conjecture that light might be bent by passing near a star because you believe that the gravity of the star will attract the light beam. If you did not specify how much, you might think your theory was confirmed by Eddington's observations. We would have no way to test between your theory and Einstein's, however, because Einstein's theory resulted in an actual number and yours did not. Einstein's is therefore the better theory, even though yours is arguably correct.

Popper recognized that you can almost always patch up a theory by employing an *ad hoc* hypothesis to explain away any contradictory result. You could probably save almost any theory with enough *ad hoc* hypotheses. Popper notes that some "genuinely testable [falsifiable] theories," even when found to be false, are nevertheless propped up by their "admirers," who use a stratagem such as introducing an *ad hoc* hypothesis or reinterpreting the theory in such a way that it cannot be refuted. At a minimum, this reduces their claim to objectivity. This is so because the *ad hoc* hypothesis itself must be falsifiable. A truly useful *ad hoc* hypothesis will have a more general purpose than simply to account for a single anomaly; if it leads to a new or improved theory, then that theory must also be falsifiable.

If the hypothesis is not falsifiable or if it leaves your original theory unfalsifiable, then your theory is not a valid scientific theory. Indeed, a theory that can explain everything is often so imprecise as to be worthless. As Friedlander [1995] notes, a theory that could not be refuted by any conceivable event is not a scientific theory. Irrefutability is not a virtue, continues Friedlander, but a vice, since a theory that is inherently irrefutable is also not testable. Thus, a scientist requires enough judgement to know when to drop a theory; to this extent, at least, the scientific method is less cut and dried than I may have implied earlier.

* * *

Popper considered astrology to be a belief system that is not falsifiable, even though it is empirically based. Indeed, he notes that rationalists have attacked astrology for its assertion that heavenly bodies could influence terrestrial events. The astrologers' assertion, however, is correct; both the moon and the sun are heavenly bodies, and they influence the tides on the earth. Do the heavenly bodies influence anything else, such as our futures? Popper argues that astrologers make vague predictions, explain away their failures, and, in that way, make their theory untestable.

Most physical scientists reject astrology because they can see no mechanism whereby the positions of the heavenly bodies can affect actions of humans on earth. The nearest bodies, the sun and the moon, can cause oceanic tides, but these are unimpressive in the sense that the oceans are several kilometers deep, whereas the tides rise and fall only a few meters above and below the mean water line. The tide does not depend on the value of the gravitational attraction of the moon on the water, but rather on the *difference* between the gravitational attraction at the surface of the water and at the bottom. Partly for this reason, shallow bodies of water have virtually immeasureable tides. The difference of the moon's attraction between the top of your head and your feet is similarly immeasurable, even if the moon is directly over your head, so it is very unlikely that a tidal force could affect your behavior. Objects, such as stars or planets, that are more distant than the moon or smaller than the sun influence even the tides in the deeper ocean immeasurably.

Additionally, according to the French astronomers Philippe Zarka and François Biraud, when astrologers construct their charts, they use obsolete data. [Henarejos, 1999] Owing to the slow rotation of the earth's axis, or *precession of the equinoxes*, the alignment of the sun, moon, and planets among the constellations gradually drifts with respect to the seasons. Orion, for example, was a summer constellation 2000 years ago. When astrologers construct your birth chart, they use the alignment of the planets as it appeared on your birthday 2000 years ago. On the other hand, when they say we are entering the age of Aquarius, they use the alignment of the planets as it is today. They can't have it both ways: "They have to choose!" say Zarka and Biraud. But they do not.

An astrologer's response to these arguments is that astrology has been empirically demonstrated to work, so there must be something missing from my theory. But has it?

Geoffrey Dean [1986-1987, 1987] is a former scientist with CSIRO, the Australian equivalent of the U.S. National Institute of Standards and Technology or the National Research Council of Canada. Unlike many critics of astrology, he has studied it in detail and is sympathetic. Dean does not consider newspaper astrology columns to be "serious" astrology, so he set out to find serious astrologers; he claims that there are about as many serious astrologers as there are psychologists. The comparison is apt, since Dean argues that a majority of Western astrologers concentrate on psychology and counseling, not prediction, and he cites many studies that show astrology's inability to predict.

How well does astrology perform as a counseling tool? Dean quotes one astrologer who discusses a meek person who has five planets in Aries. Having planets in Aries is supposed to signify aggression, and having all five planets in Aries is very unusual, so the meek person ought to be very aggressive. Thus, the astrologer looks for other signs that indicate meekness. That is, he looks for an *ad hoc* hypothesis with which to explain away his failed prediction. If he cannot find such signs, he employs another *ad hoc* hypothesis, specifically, the hypothesis that sometimes the opposite of what he predicts also happens:

> [I]f a person has an *excess* of planets in a particular sign, he will tend to suppress the characteristics of that sign because he is scared that, if he reveals them, he will carry them to excess. But if on the next day I meet a very aggressive person who has five planets in Aries, I will change my tune: I will say that he *had* to be like that because of his planets in Aries. [italics in original]

In other words, if you have five planets in Aries, then, as a direct result, you will be either aggressive or not aggressive.

35

Since everyone is either aggressive or not aggressive, or sometimes aggressive and sometimes not aggressive, the prediction is necessarily accurate, but it also is not testable and therefore valueless. Indeed, Dean quotes one astrologer, who recognizes that astrology cannot in reality make credible evaluations, as saying, "Any good I've done as a consultant, and I have done some good, had less to do with my being a good astrologer than with my being a good person." Dean therefore considers astrology to be potentially useful, even if it is completely untrue. Unfortunately, he does not evaluate the astrologer's statement that he has "done some good," a claim for which we have only the astrologer's word and no independent evaluation.

To demonstrate that astrology is, in his word, untrue, Dean describes a number of tests, such as giving right and wrong charts to subjects and asking them to pick which is right. That is, astrologers whom Dean considers serious are asked to evaluate two charts for each subject: the correct birth chart and also the chart of someone else or a chart in which signs have deliberately been changed or reversed. Each subject is then asked which of the two evaluations more accurately describes him or her. In no study cited by Dean could subjects discern their own charts more often than would have occurred by chance. Dean notes, by contrast, that subjects can recognize themselves with greater probability in valid personality tests, so the problem is with the astrology, not with the subjects or the methodology. Dean concludes that astrology is not real but can still help people as long as it "satisfies" them. But he provides no evidence that astrology is effective even in this limited way.

* * *

Astrologers, you may think, are an easy mark. But people who are more widely considered scientific also develop nonfalsifiable theories. David Stannard, in his book *Shrinking History*, [1980] gives an example of a supposedly scientific

36

theory that is also not falsifiable: psychoanalysis. (By psychoanalysis, I mean Freudian psychoanalysis as such, not psychotherapy in general.) Stannard, a historian, was Professor of American Studies at the University of Hawaii when he wrote *Shrinking History*. His book is an examination of a subdiscipline called psychohistory, which is based on psychoanalysis, so Stannard devotes much of the book to a study of psychoanalysis in general. Stannard sees psychoanalysis as a theory that is not falsifiable and therefore not testable.

Stannard gives the example of what psychoanalysts call the anal-erotic character. People who are anal-erotic are supposedly frugal, obstinate, and orderly, perhaps to a fault. In addition, they are supposed to be slightly sadistic. They get that way because of their toilet training. Sometimes, however, anal-erotics repress their inherent sadism and behave in the opposite way; this is called reaction formation. Reaction formation is an *ad hoc* hypothesis designed, like the astrologer's *ad hoc* hypothesis, to explain away a failed prediction. Thus, as Stannard puts it, a person who is anal-erotic may be frugal or generous, orderly or disorderly, sadistic or kind. In other words, I can safely diagnose anyone as anal-erotic without fear of contradiction. Such a diagnosis has no predictive power and cannot be falsified. It is completely worthless.

Stannard goes on to explain that defense mechanisms such as repression or denial can be used similarly to allow diagnoses for which there is not a shred of direct evidence. Indeed, sometimes your denying certain charges may be taken as further evidence of the truth of the charges, as in cases of alleged child abuse or false memory syndrome.

Does psychoanalysis nevertheless succeed as a therapy? Psychoanalysts generally resist evaluation by independent observers, so it is difficult to tell. We can, however, get some evidence from their reactions to an article by Frederick Crews in *The New York Review of Books*. Crews is a former chairman of the Department of English at Berkeley and a persistent critic of psychoanalysis. His 1993 review of several books critical of

37

Freud and psychoanalysis caused a furor, at least among practicing psychoanalysts. That article, a later article on false memory syndrome, and the responses of Crews's critics were compiled, along with an afterword, in the book *The Memory Wars: Freud's Legacy in Dispute.* [Crews, 1996] Crews leaves no doubt where he stands: He considers psychoanalysis a dangerous pseudoscience.

Mortimer Ostow is a practicing psychiatrist and psychoanalyst, and Professor Emeritus of Pastoral Psychiatry at the Jewish Theological Seminary. In response to Crews, he reviews 37 of his own cases. Of these, Ostow says, 14 percent exhibited "dramatic" improvement, as witnessed by friends and relatives who were unaware that the patients were undergoing therapy. Another 25 percent considered themselves to have improved "impressively." Thus, 40 percent of the group reported marked improvement. Ostow admits, almost as an afterthought, that he used medication in almost half of his sample but does not specify which patients. He claims that the changes he sought in evaluating the outcomes were personality changes and not simple changes of mood induced by drugs. He considers the results "not at all discouraging" considering the nature of the illnesses and their durations.

Ostow seems completely unaware of proper scientific methodology. He bases his evaluations entirely on anecdotal evidence and on his own observations. He seems unaware of the problem of experimenter bias and, precisely like the astrologer who has "done some good," assigns to himself a dual role: therapist and evaluator. This is precisely the problem that Maddox's team observed in Benveniste's laboratory.

Psychoanalysts like Ostow appear more likely than other therapists to suffer from such experimenter bias. Stannard [1980] cites a study that was based on fictitious transcripts of therapy sessions alternately labeled "early" and "late"; it showed that psychoanalysts were more likely than other therapists to perceive positive change where none existed. Similarly, psychoanalysts were likely to perceive interviewees as more disturbed than were other therapists, especially when the interviewees were presented as patients, rather than as job applicants.

Ostow ignores the confounding effects of the medication on the evaluations except to remark that he ruled these effects out on the basis that the medications cannot produce the changes he observed: an *ad hoc* hypothesis. He uses no control group and seems as unaware of the placebo effect as of experimenter bias. Crews considers Ostow's failure to control for other variables "pathetic" and argues that, as a result, Ostow's survey wholly lacks validity.

Lester Luborsky, Professor of Psychology and Psychiatry at the University of Pennsylvania, writes that Crews is mistaken when he claims that psychoanalysis has not been evaluated as a therapy. Luborsky has compared psychoanalysis with other therapies and has found that it is "at least as good as other forms of psychotherapy." In 1980, when Congress was debating whether to include psychotherapy under Medicare, they sought proof that it was effective. Estimates then were that there were 100 to 140 different schools of psychotherapy. [Marshall, 1980] Should not a truly valid therapy stand out above most of the others? Luborsky tells us only that psychoanalysis is not so bad by comparison.

Besides damning with this faint praise, Luborsky shows that he is unaware of the concept of falsifiability. A theory that is no worse than any others is also no better, or at least not demonstrably better. If psychoanalysis is a valid scientific discipline, then it must make precise predictions, and these predictions must be more accurate than the predictions of competing theories. Furthermore, according to Crews, Luborsky has incorrectly compared psychoanalysis with receiving no treatment, whereas he should have compared it with a placebo.

Stannard cites a study that did precisely that: compared psychoanalytically oriented treatment with behavioral therapy and with being placed on the waiting list. The wait-listed group was considered a control, and we may regard this group as having received a placebo; these patients were frequently reassured that they had not been forgotten and would soon be offered treatment. The patients in the two therapy groups advanced slightly faster than the control group, but half of the control group also showed significant improvement. The outcome of the study was evaluated by three groups: the patients, the therapists, and a group of outside evaluators who did not know into which category each patient had been placed. The evaluations of the patients and their therapists may well suffer from experimenter bias; the judgement of the outside evaluators is therefore probably the most nearly unbiased. According to Stannard, if we consider only the outside evaluators' judgement, then all three groups improved dramatically, but there was no significant difference among any of them. Ostow's 40 percent success rate and Luborsky's faint praise merely underscore that psychoanalysis does not work. It is a pseudoscience that, like astrology, is set up to be untestable. When it is tested anyway, it fails.

* * *

By contrast, astronomers have found that galaxies rotate about their centers faster than they "should" according to current theory. Roger Welsch, [1994] a folklorist who writes a monthly column in *Natural History*, claims that astronomers simply "readjust the uncooperative universe to fit their theories," invent dark matter (an *ad hoc* hypothesis), and "just like that, the modern, popular theories are back in business." Now, I know that Welsch is trying to appear funny. Nevertheless, he fosters a serious misconception; in fact, cosmologists have *postulated* missing mass and then set out to see *whether* they could find it. If the missing mass can be found, it may or may not be able to explain the observations.

Astrophysicists have identified candidates for the missing mass. One possibility is that certain particles called *neutrinos* carry some of the mass. Conventional theory considers neutrinos

40

to be massless, but some astrophysicists have postulated that they have a small but finite mass. Attempts to measure the mass of the neutrino have resulted in evidence that they are not wholly massless. Similarly, astrophysicists have postulated the existence of heavy particles called WIMP's, for *w*eakly *i*nteracting *m*assive *p*articles, but there is so far no evidence for WIMP's.

A third candidate for the missing mass is small planetoids that are similar to Jupiter but float around in the cosmos not bound to any star; in effect, they are stars that are too small to "ignite." As I write, such planetoids have been discovered, but it is not clear whether or not they have enough mass. Thus, cosmologists are still looking, and the theory is not yet "saved." Welsch notwithstanding, it will not be saved unless there is evidence to save it. In short, the astronomers, unlike astrologers and psychoanalysts, have developed an *ad hoc* hypothesis that is testable and are testing it. This *ad hoc* hypothesis is not a patch designed to save a failed theory.

The discovery of the neutrino itself is pertinent to our topic. Physicists believe strongly in the "law" of conservation of energy. Galileo first recognized that the product of the mass of a particle and the square of its velocity had a special significance. Later, one-half that product was recognized as the *kinetic energy* of the particle. Over the years, physicists realized that there were other forms of energy and postulated that energy can neither be created nor destroyed; rather it is *conserved*, or transformed from one form to another. This is called the Law of Conservation of Energy, even though it is really a postulate. No violation of this law has ever been found, however.

When the Austrian physicist Wolfgang Pauli looked at a certain nuclear reaction and found that energy appeared not to have been conserved, he postulated that the missing energy was carried off by a new particle that we now call a neutrino. He next developed a mathematical model to predict certain properties of neutrinos, so that their existence could be verified. More than 25 years later, in 1956, Frederick Reines and Clyde

41

Cowan detected the first neutrinos; now neutrinos are well-established if still hard to detect. Had neutrinos not been detected or had they been shown to have the "wrong" properties, then Pauli's theory would have been falsified. Unlike Benveniste, Rubik, and Ostow, scientists would have accepted the verdict of reality and sought another explanation.

The parallel with the missing mass problem is striking. When faced with an anomaly in a well-established theory, scientists do not exactly patch up the theory (as Welsch wrongly implies) but rather seek an explanation within the realm of the theory *and search for evidence* to support or refute their hypothesis. More accurately, what may appear to be a patch is in fact a precise and specific hypothesis that is then subjected to further testing. That is, where astrologers and psychoanalysts think that it is enough to merely state an *ad hoc* hypothesis, the astronomers Welsch castigates propose precise and testable hypotheses and subject them to potentially withering scrutiny. If a hypothesis withstands the scrutiny, then the theory has been extended or improved. If no hypothesis succeeds, then even a well-established theory may ultimately fall or require major revision.

* * *

We are left with the question, What does it take to refute a hypothesis? That is, when faced with a bit of unexplained evidence, when is it legitimate to explain it away or ignore it? The orbit of Mercury, for example, is not a stable ellipse. Rather, Mercury's perihelion, the point at which it passes closest to the sun, revolves very slowly about the sun. This is called the precession of the perihelion of Mercury. Because of the precession, Mercury's orbit is not quite closed, not quite an ellipse. Most of the precession can be attributed to the gravitational pull of the other planets, which distorts the orbit slightly. But careful calculations based on Newton's laws leave a small component unexplained. This small component was a serious problem that Newtonian mechanics could not explain, and it was not explained until Einstein proposed the General Theory of Relativity in 1916.

Before the development of General Relativity, a skeptic might have pointed to the precession of the perihelion of Mercury and argued that Newtonian mechanics was simply wrong. Newtonian mechanics, however, worked well on virtually every other solvable astronomical problem to which it had been applied. Most scientists, therefore, regarded the precession as an anomaly or a minor annoyance that bore explaining but did not falsify Newtonian mechanics as a whole. Indeed, anyone who argued that Newtonian mechanics was wrong because of a single small anomaly would probably have been regarded as a crank, since the preponderance of the evidence supported the theory. By contrast, if an astronomer had discovered a single asteroid going about the sun in a triangular orbit, Newtonian mechanics would have been conclusively falsified.

Today, we would probably say, somewhat circularly, that Newtonian mechanics is correct within the domain within which it works, but it has been encompassed by a later theory that has a wider domain of applicability. Note, however, the difference between this statement and the statement that Therapeutic Touch works on those patients on whom it works. We can predict in advance precisely where Newtonian mechanics will work and where it will not. For example, if a particle's velocity is 10 percent of the velocity of light, then its kinetic energy is about 1 percent more than the value predicted by Newtonian mechanics. For many purposes, then, we can make the specific statement that Newtonian mechanics works as long as a particle's velocity is less than 10 percent of the velocity of light. Therapeutic Touch or any other fringe treatment, by contrast, makes no predictions as to its range of validity and will be declared to have worked on certain patients only after the treatment has been applied.

43

Rational, irrational, and nonrational

Popper argues that a theory that is set up so that it can never be disproved is not falsifiable and therefore not scientific. I would go further. I argue that any beliefs about the character of the external world must be falsifiable in order to have value. Specifically, we must regard all our beliefs more as hypotheses to be tested than as beliefs to be defended. Here, by beliefs about the character of the external world, I mean claims of fact, not beliefs or opinions about morality or ethics.

I once attended a seminar at which a rabbi (either Jacob Staub or David Teutsch of the Jewish Reconstructionist Federation) made an important distinction between *ir*rational and *non*rational. The distinction is easily elucidated by example. For instance, the belief that all people are created equal is not a rational belief in the sense that you can neither prove nor disprove it. It is, however, not irrational to hold such a belief. That belief is nonrational. So are the belief that people should be slaves or pawns of the state and the belief that slavery is wrong; the belief that people are entitled only to what they can earn from the marketplace and the belief that people are entitled to a minimum standard of living. Though I find some of these beliefs distasteful, I cannot see any way to adduce evidence for or against any one of them. Such beliefs are therefore not irrational in the sense that they fly in the face of evidence; they are, instead, nonrational. On the other hand, the belief that the earth is flat or that the moon is made of green cheese is contrary to known fact and is therefore irrational. Less obviously, so is the belief that the universe is approximately 10,000 years old.

The same rabbi pointed out, by the way, that what is objectively rational or irrational may change with time. For example, he noted that the medieval rabbi and philosopher Moses Maimonides argued that a man could never fly. I assume that Maimonides meant "fly by flapping his arms." I do not know whether anyone in his time could have conceived of a machine for flying; I assume, however, that Maimonides would

44

also have denied the possibility that we could ever build a machine—a flying boat?—in which a man could fly. At the time, that would have a perfectly rational thing to believe, even though we now know it is not true. If you believed that today, however, we would be justified in calling you irrational. In other words, what is rational at one time may become irrational as we learn more about the world.

Self correction

Science typically corrects its own mistakes. As a rule, you can't falsify data or report inaccurate conclusions and succeed for very long. If someone performs an experiment and reports it in the literature, it is likely to be replicated elsewhere. If the experiment has been done correctly and interpreted correctly, then its conclusions will hold up. If not, then the conclusions will ultimately be modified or rejected.

Consider the case of polywater. [Friedlander, 1995] This was a form of water that seemed to have different properties from normal liquid water: different boiling point, different freezing point, different viscosity. It was first reported in the Soviet Union and was introduced to the West by Boris Deryagin of the Institute of Physical Chemistry in Moscow. According to Friedlander, Deryagin was a respectable scientist and the winner of a prize that was roughly equivalent to the National Medal of Science in the U. S. If Western scientists were skeptical, says Friedlander, their skepticism was tempered by Deryagin's reputation.

The molecules in a liquid are held together somewhat loosely by electrical forces. They are not bound as tightly as those in a solid but, rather, are comparatively free to move around; this is what gives a liquid its fluidity. Polywater behaved differently on certain surfaces than ordinary water, and scientists at the National Bureau of Standards suggested that it might be a polymerized form of water. That is, they suggested that the water molecules may have been bound together in long

45

chains, like those in a plastic. If the chains were loosely bound to each other, the polymerized water would still be a liquid but would have higher viscosity and a higher boiling point than ordinary water. Additionally, the NBS scientists presented spectroscopic evidence that supported their suggestion. So many theories were proposed that Friedlander warns, "There are many more wrong answers than right ones, and they are easier to find."

Polywater could be prepared only in fine silica or glass capillaries; attempts to produce it in large quantities failed. Silica is one of the main components of common glasses and is slightly soluble in water. When other experimenters found impurities, including silica, in their samples of polywater, Deryagin explained that it was necessary to keep everything extremely clean. Eventually, however, as evidence for impurities mounted, even Deryagin recanted and agreed that polywater was not a new form of water, but rather the result of contamination. [Levi, 1973]

* * *

The more recent sensation about cold fusion is another example. Although some say the jury is still out, it now seems almost certain that cold fusion was an error. [Friedlander, 1995]

In 1989, Stanley Pons and Martin Fleischmann, both well respected electrochemists at the University of Utah, thought that they had discovered a new and unexpected phenomenon that they called cold fusion. They submerged an electrode of palladium in a flask of heavy water. This is water composed of the deuterium isotope of ordinary hydrogen; deuterium is identical to hydrogen, except that its nucleus holds one neutron in addition to one proton. If you can get two deuterium nuclei to fuse together, you can derive energy from the process. This is how the hydrogen bomb works, and plasma physicists have dreamed for decades of being able to develop a *controlled* fusion reaction for generating electrical power. Their experiments with tokamaks and other fusion reactors cost billions of dollars.

46

The Utah researchers thought they saw fusion in a bottle. In essence, they performed an electrolysis experiment, similar to what you do when you pass an electric current through a salt-water solution, dissociate the water molecules, and liberate hydrogen and oxygen. When Pons and Fleischmann did the experiment with palladium electrodes and heavy water, they thought that they detected more energy than they "should" have, that is, more energy than they could explain by chemical reaction. They attributed the excess energy to fusion of deuterium ions in the electrode. They also saw gamma rays, neutrons, and small amounts of tritium, a form of hydrogen with one proton and two neutrons. Friedlander says these are precisely what you would expect to see if fusion of the deuterium atoms had taken place, but the quantities Pons and Fleischmann observed were within the range of normal background. Nevertheless, Pons and Fleischmann's claims were made plausible by closely related experiments by Steven Jones at Brigham Young University.

Laboratories rushed to replicate the Utah experiments. Ultimately, they could not be replicated reliably, and very few laboratories are now investigating cold fusion. It is uncertain precisely why Pons and Fleischmann saw what they did, but it seems likely to have been a very public mistake. Within a year, however, it was recognized as such.

Polywater and cold fusion are two examples of science correcting its mistakes. I have not read much about Pons and Fleischmann after the discrediting of cold fusion, but note Deryagin's initial reaction: to say that everyone else was not careful enough, everyone else's polywater was contaminated. Unlike others we have discussed, however, Deryagin did not hold unreasonably to his position, but rather conceded his error. It is for this reason—because serious scientific errors can be made apparent—that I argue that science is more to be trusted than virtually any other intellectual endeavor.

Paradigm shifts

Thomas Kuhn, a physicist who later turned to the history and then the philosophy of science, taught at Harvard, Berkeley, and Princeton, and was associated with the Institute for Advanced Study in Princeton. In his influential book *The Structure of Scientific Revolutions*, [1970] Kuhn argues that science does not advance cumulatively but rather by jumps. He calls these jumps *revolutions* and calls the periods between revolutions periods of *normal science*. A revolution can be a global revolution that involves virtually the whole world, like the Copernican revolution, or it can be a revolution that takes place in a very narrow field. In either case, during a period of normal science, the practitioners of a given field subscribe to a *paradigm* or a set of paradigms. As Kuhn acknowledges in the postscript to his second edition, he uses the word paradigm ambiguously. Sometimes, he means the entire constellation of beliefs, values, problems, and problem-solving techniques that are shared and used by the community. At other times, he means rather an archetype or example that scientists use to solve problems that are in some ways similar to the archetype; archetype is closer to the dictionary definition. For our purposes, however, we may adopt the first meaning: the entire worldview of the scientists in a given field.

Kuhn argues that, during "normal" times, scientists adhere to a given paradigm and use it successfully to solve problems. The existence of a single, uncontested paradigm is what defines normal science, and adherence to a shared paradigm is part of what defines a specific field. Sometimes, however, scientists discover an *anomaly*, that is, an experimental fact that their paradigm cannot account for. During the *crisis* that follows, some scientists will propose daring new theories and new paradigms. Two or more paradigms will therefore compete for the allegiance of the community. Eventually, one of those paradigms will prevail, and a period of normal science will

48

follow. That is, periods of normality are punctuated by periods of revolution. During the revolution, fundamental concepts and theories are challenged and ultimately replaced by new concepts and theories. After the revolution, normal science begins again but with a new paradigm, that is, with new concepts and theories.

Kuhn is at pains to point out that the change to a new paradigm is not irrational, but neither are those who refuse to accept the new paradigm necessarily illogical. He denies that Popper's idea of falsifiability applies during the period of crisis, but says rather that rejection of the old paradigm is based on its inability to account for as many experimental facts as the new paradigm. Thus, the old paradigm is not falsified but rather is rejected on the basis of a preponderance of the evidence. Neither the old nor the new paradigm, however, can account for all the evidence, so neither can be ruled in or out by simple logical argument. Rather, a period of persuasion follows, and often the community is almost wholly converted to the new paradigm. This is sometimes called a *paradigm shift*.

Kuhn's work has therefore been interpreted as evidence that scientists do not change their collective mind when presented with compelling new evidence but rather when their culture or their worldview changes, rather as composers stop composing classical music and start composing romantic music or artists stop producing impressionist paintings and start producing cubist paintings. The analogy is striking: Impressionism, once itself thought ugly, becomes the paradigm. A few painters start producing cubist or other "modern" paintings. Like impressionism before it, modern painting is not well received. Slowly, a few painters accept it and defend it, then more, and finally it is generally accepted by the community, and no one paints impressionist paintings any more. The change from impressionism to modern painting is a change of paradigm. The direction in which painting went, however, was arbitrary in the sense that there was no compelling physical reason to adopt modern painting. Acceptance of modern painting was purely a matter of taste, and you are free to accept it or reject it. The same is not true of well-established scientific theories; you may be free to like or dislike them, but you are not free to accept or reject them arbitrarily.

The mathematician John Casti [1989] compares scientific knowledge to an unknown continent. At first, we get a few sketchy maps of the coastline, and we have to guess at the rest. We draw sketchy maps with, if we are honest, large blanks where we know nothing. After a while, one map is deemed to be more accurate than the others, perhaps because it has more recent data than the others or because several explorers have returned with similar accounts. That map then becomes our paradigm for describing the new continent. Later, additional explorers return with newer information and publish their own maps. Gradually, these maps replace the earlier map and become the new paradigm. And so it goes, until we develop a map that is so accurate that we can use it unerringly to navigate from one place to another.

It is important to remember that the map is not the continent. But Casti fails to draw what is to me the obvious conclusion: The maps are based on objective reality, and they describe the continent with ever-increasing accuracy. In the same way, *scientific paradigms are based on objective reality, and they describe that reality with ever-increasing accuracy.*

Kuhn's idea of a paradigm shift has nevertheless been misunderstood to mean that science is arbitrary in that scientists are not governed by reason or evidence but rather believe whatever the dominant culture prescribes. Kuhn argues the contrary. Painting is a matter of taste, but the change of scientific paradigms is based on evidence such as the ability to solve certain kinds of problems. The new paradigm is not predictable in advance, and another paradigm might do as well, but any paradigm whatsoever will not do at all. If Einstein had not developed the General Theory of Relativity, someone else would have developed another theory of gravitation. It may not have worked as well as Einstein's formulation, or it may have worked better. But it would not have been accepted if its proponents had not convinced the larger community that it accounted for the observed facts better than Newton's theory.

* * *

Kuhn based much of his thinking on the Copernican revolution: the shift from the old geocentric universe of Ptolemy to the new heliocentric universe of Copernicus. Before Copernicus, most people, including scientists and astronomers, believed that the sun, the moon, the planets, and the fixed stars revolved around the earth. Copernicus, however, postulated that the earth and the planets revolved around the sun. This was a great simplification and made calculations of orbits easier, but in fact there was little evidence to support one view over the other.

Copernicus lived at the very dawn of modern science, and the Copernican revolution was in some ways the greatest intellectual revolution in history. Before Copernicus, everything revolved around the earth. The view was consistent with religious beliefs of the time, and it was sometimes physically dangerous to espouse the heliocentric theory. Nevertheless, after Copernicus, everything was changed, and we could no longer regard ourselves as the center of the universe. Eventually, we found that we were not at the center of the galaxy either, and that the galaxy was only one among billions.

The Copernican theory was not immediately and universally adopted. Furthermore, the Ptolemaic theory was not exactly wrong; it was always possible to view the earth as the center and everything else as revolving about the earth. You can choose the center of your coordinate system anywhere you like: Times Square, the center of the earth, the center of the sun. The problem with the Ptolemaic theory was that the choice of the earth as the center was inconvenient (and would have made finding an accurate force law virtually impossible), not that it was wrong. Convenience is an important criterion, however, and, as more-accurate observations became available, the predictions of Ptolemaic theory became less and less accurate.

51

This was so because the planets' orbits about the sun are elliptical, not circular. The Ptolemaic theory tacitly assumed circular orbits, but transformed them to a coordinate system whose center was the center of the earth. Arab astronomers tried to patch up the theory but without success. If they had been able to assume the equivalent of elliptical orbits, however, the theory would have predicted the positions of the planets accurately. They could not have adopted elliptical orbits very easily in the Ptolemaic coordinate system, but it was relatively simple in the Copernican system. Thus, the Copernican system triumphed at least in part because it was simple and convenient, but not because it was right. (In fact, it is not right, in that the sun is the center of the solar system but not the center of the universe, as Copernicus thought.) Kuhn calls adopting the Copernican theory and discarding the Ptolemaic theory a change from one paradigm to another. The new paradigm was adopted not because it was necessarily better nor closer to the truth but because it solved certain problems better than the old paradigm.

* * *

By contrast to the Copernican revolution, let us consider another revolution: one less well known and less far-reaching but nevertheless a real revolution in geology: continental drift. [Friedlander, 1995] Alfred Wegener proposed this theory in 1912, at a time when geologists considered the earth's surface solid and immutable. Wegener had noticed that the coastlines of Western Europe and Africa were a mirror image of the coastlines of the Americas. (So did we, when I was in grade school. We were told not to be silly. It was not the only time we were told not to question some dogma or another.) Wegener next compared the continental shelves and found them to be better fits than the coastlines, which had no doubt been distorted by erosion. He analyzed the elevations of the earth's surface and showed that there were on the whole two elevations, the elevation of the land and the elevation of the sea floor, with only relatively minor variations from those elevations. If the earth had simply cooled from a molten sphere, you would expect a wrinkled sphere with a single average elevation and relatively slight variations from that elevation. You would not expect two discrete elevations. Wegener further noted that the longitude of a certain island near Greenland had apparently shifted about 1 kilometer westward between 1823 and 1907.

Wegener also enlisted the well known anomaly that fossils of related species were found across the Southern hemisphere but not in the Northern, for example, in South America and Africa but not Europe or North America. It was unreasonable to assume that these species had once existed at all points in Northern Africa, Europe, and North America, but had left no trace, so geologists proposed that there had once been "land bridges" between Africa and South America. Wegener argued that these land bridges could not have disappeared because, like the continents, they must have floated on a denser interior and so could not have disappeared any more than a continent can disappear.

In spite of the wealth of evidence in its favor, Wegener's theory was not generally accepted until the 1960's, when evidence of material welling up from the ocean floor was discovered. This evidence was based on weak magnetic fields frozen into the rock on the ocean floor. [Friedlander, 1995] That is, a magnetic map of the floor of the Atlantic Ocean turned out to be "striped," and the frozen magnetic fields were oriented in opposite directions in adjacent stripes. Furthermore, the pattern of the stripes was symmetrical about a center line that runs north to south and roughly bisects the Atlantic Ocean. This was interpreted as evidence that molten rock was constantly welling up from the earth's interior and spreading out from the center line as the continents drifted apart. The earth's magnetic field magnetized the rock; when the rock hardened, the magnetization of the rock was frozen in place. As the earth's magnetic field changed direction over geological times, the magnetization of the rock also changed direction. This evidence was considered

53

definitive and allowed Wegener's theory to be accepted relatively rapidly.

You could argue that geologists should have been faster to accept Wegener's theory, that their reluctance establishes the arbitrariness of the prevailing paradigm. In their defense, however, I would counter that there are a lot of crackpot theories, and sometimes it is hard to recognize an original but bizarre contribution. Also, Wegener was a professional meteorologist, not a geologist, so it was easy to dismiss him. And there were serious objections: In particular, where did the force to move the continents come from? Calculations suggested that drift was not possible, though Arthur Holmes suggested, correctly, that convection currents in the molten core of the earth provided the required force. In any event, when the evidence was virtually irrefutable—a smoking gun—the theory was adopted almost overnight: on the weight of the evidence. Once the crucial bit of evidence was recorded, the revolution took less than a decade.

* * *

Kuhn and others believe that it is easier to convince young scientists than older, and that the new paradigm is adopted only when the old have died off and the young have taken over. Kuhn quotes Max Planck, the founder of the quantum theory, as saying a new scientific truth does not triumph by convincing its opponents and making them see the light, but rather because its opponents eventually die, and a new generation grows up that is familiar with it.

Ernst Mach (as in Mach number) admitted in the early 1900's that relativity and even the atomic theory are too much for him. In the Introduction to my English translation of *The Science of Mechanics* [Mach, 1960], the translator Karl Menger quotes Mach as saying,

I do not consider the Newtonian principles as completed and perfect; yet, in my old age, I can accept the Theory of Relativity just as little as I can accept the existence of atoms and other such dogma.

I consider this remark especially ironic, because Einstein based much of his thinking on what is now called Mach's principle.

If Planck were right, it would argue in favor of taste or dogma rather than reason as the driving force behind acceptance of the new paradigm. If you cannot convince older, presumably more expert scientists, then how is reason behind the change of paradigm? Scientists, in this view, merely adhere to the dogma in which they have been indoctrinated or have a vested interest, and the younger scientists have been indoctrinated in the latest dogma, whereas the older have not. What is not clear in this argument is precisely who indoctrinates the younger scientists if their teachers have not generally accepted the latest dogma.

David L. Hull, Professor of Philosophy at the University of Wisconsin, and his colleagues tested what they call "Planck's principle" by examining the ages at which prominent scientists accepted the theory of evolution. [Hull et al., 1978] Because "Darwinism" can have different meanings to different people, they circumscribed their study by looking only at the date at which the scientists in the study were known to have accepted evolution. They did not have to concern themselves with whether these scientists believed in the heritability of acquired characteristics or in natural selection, for example. They studied only those who were at least 20 years old when Darwin's *Origin of Species* was published; scientists who were substantially younger than 20 could well be assumed to have been taught Darwinism as part of their training. Similarly, they restricted their study to British scientists, partly because Darwin was British and the data for Britain were most readily available.

Hull and his colleagues studied the 10 year period following the publication of the *Origin of Species* in 1859. They tabulated

55

the age in 1859 of a sample of 67 British scientists, the date of their acceptance of evolution (if any), and the age in 1869 of the holdouts. By that date, only about 75 percent of the scientists in the sample had accepted evolution.

The researchers performed three tests: First, they calculated the average age in 1859 of those who had accepted evolution within 10 years and the average age of those who had not. The average age of the accepters was 40, whereas that of the rejecters was 48, so they conclude that age was a factor in accepting evolution, at least within a 10 year period. Additionally, they performed a statistical analysis that suggests that less than 10 percent of the variation in acceptance is the result of age. Finally, they calculated a statistic that measures the average speed with which the accepters accepted evolution. There was no significant variation with age; that is, among scientists who changed their minds, the older scientists did so as quickly as the younger. Planck and Kuhn notwithstanding, it appears, therefore, that age was not the major factor that determined who adopted the new paradigm and who did not. Hull and his colleagues conclude that "reason, argument, and evidence" must explain at least some of the acceptances.

* * *

Thus, whereas I certainly agree that we are all products of our cultures, I argue that science—at least the physical sciences, including molecular biology—does not flit randomly from paradigm to paradigm (or worldview to worldview) but rather that

> *Science progresses, sometimes rapidly, sometimes incrementally, toward some kind of objective description of reality.*

I mean that statement in this sense: The philosopher Ian Barbour [1974] considers scientific models to lie somewhere between true pictures of reality and metaphors. They are not literal pictures of reality, but neither are they merely fictions, in the same sense that a map is not a literal picture but is still a representation of a geographic area. Barbour notes that the ancient Babylonians kept careful records and were able to predict solar eclipses accurately simply by studying their records. They had, however, not the slightest idea what caused a solar eclipse. That is, they had no model. Today, by contrast, we think of the sun, the moon, and the earth as spheres that orbit one another in a certain way. That is our model, and with it we can explain how the moon occasionally but very predictably interposes itself between the sun and the earth to cause an eclipse. Similarly, by pretending that gas molecules are hard spheres, we can derive the gas laws, which are otherwise merely experimental facts with no underlying explanation. These models truly describe reality, even though planets, stars, and gas molecules are not truly spheres. Models are hard to develop, and scientists rarely have to choose between more than two models. Even though the choice is sometimes difficult and involves controversy, it is by no means arbitrary.

True, I do not know exactly what an electron is, and I do not understand wave-particle duality, but I know the mass and the charge of the electron with an uncertainty less than 1 part per million. I know a quantity called the g-factor of the electron with an uncertainty of 1 part in 100 billion. These are numbers that will probably never change, but, at any rate, they are independent of the culture or belief system of the person who measures them: They cannot be subjected to revisionism or, as some would say, cannot be deconstructed.

B. F. Skinner, the psychologist and developer of behaviorism, complains in his book *Beyond Freedom and Dignity* [1972] that the physical sciences have matured greatly since the ancient Greeks, but their psychology still makes sense today. He makes an essential point: Physics, chemistry, molecular biology, and geology are mature sciences. Their practitioners have arrived at almost universal consensus because they study real, verifiable phenomena. This consensus is not an accident brought about by indoctrination of the young but rather is an inevitable consequence of weeding out bad theories by careful study and experimentation.

Practically no reputable scientist disputes the existence of electrons, the theory of continental drift, the germ theory of disease, nor, I should add, geological and biological evolution. Indeed, one way to distinguish a mature science from a science that is still struggling to become objective is to note whether its practitioners generally agree or whether there are gigantic and sometimes bitter rifts, as there are today between cultural and scientific anthropologists. [Morrell, 1993] As my colleague Eric says, "The less evidence there is for a given proposition, the harder people fight over it." There is very little bitter fighting in the physical sciences.

Taste

The poet and novelist Anne Michaels [1997] has a character say, "The truth doesn't care what we think of it." Put slightly differently,

> *The truth or falsity of a scientific theory does not depend on whether or not you like its outcome.*

Indeed, that is true of any belief. But let me give a brief scientific example. Richard Dawkins has popularized a theory called the *selfish gene theory*. I will have more to say about it later (see "The biological origin of evil," Chapter 5). For now, let us just note that the selfish gene theory postulates that evolution works by survival of the fittest *gene*, not necessarily the fittest *organism*. That is, much is explained if we assume that an organism's function is to propagate its genes, as opposed to propagating its species or reproducing itself.

I once read what a dreadful theory the selfish gene theory is: It can be used to justify rape. How it can be used to *justify* rape escapes me, although it certainly may help to *explain* rape or promiscuity. For example, armed with the selfish gene theory, you can easily guess why men are more likely to be promiscuous than women: because one man can impregnate many women in a year, whereas one woman can, in general, bear only one offspring in a year. Thus, if men are promiscuous, they might be able to ensure the survival of their genes by giving them many more opportunities to propagate. Women, by contrast, can bear only one or two children a year, so they have to be choosy and pick mates who will give those children the best chance for survival. Thus, men are likely to be promiscuous; women, less likely. This is a completely amoral discussion that tries to *explain* promiscuity, not justify it. Justifying promiscuity in modern humans, if you could, would be an ethical discussion, not a scientific discussion, though, as we will see, even ethics may have a biological origin and can be subjected to scientific scrutiny.

Let us pretend, nevertheless, that the selfish gene theory could in fact be used to justify rape. That would not make the theory wrong, nor would it imply that we should not advance the theory. To the contrary, if we could use the theory to understand rape, perhaps we could eliminate it. But the writer who thought that the theory was reprehensible abhorred selfish gene theory because he did not like what he thought was one of its outcomes.

* * *

A corollary to the preceding principle might be

> *The truth or falsity of a scientific theory does not depend on whether or not you understand it.*

For example, those who refuse to believe in evolution often ask, "What is half an eye?" as if by asking they automatically ridicule the idea that an eye could have evolved from some less complex organ. The answer, however, is simple: Half an eye is

59

a light-sensitive spot or a retina. Three-quarters of an eye is a fly's eye, which has many receptors but no real lens and no focusing mechanism. A whole eye (such as a mammalian eye) is an eye with a lens, a retina, and a focusing mechanism. This is the eye that is implied in the question, but, in fact, there are many different kinds of eyes in nature.

Now, I will frankly admit that I cannot visualize the transition from one of these to another and then to an eye. Is that a failure of theory or of my imagination? There is enough evidence in favor of evolution that I conceded that the fault is with my imagination, not the theory, and went to the literature.

Richard Dawkins is Fellow of New College, Oxford, a Lecturer in Zoology at Oxford, and the author of several books about natural selection. Dawkins [1987] notes that scientists in other fields sometimes oppose evolution because they cannot imagine that the necessary development could occur within a finite time. He calls this viewpoint the Argument from Personal Incredulity, though to my mind there is also an element of Willful Ignorance. To make the evolution of the eye more credible, Dawkins suggests that we imagine something, not quite a fully developed eye, but something that could develop in a single step into an eye. What you imagine might be different from what Dawkins or I imagine. Dawkins calls it X. Now, he says, imagine something that could have developed in a single step into X; call that X'. Work your way backward through X to X' to X'' to $X^{(3)}$, and so on, until you reach something that does not in any way resemble the mammalian eye. Can the eye have developed from some very distant structure, call it $X^{(n)}$, to its present form? Dawkins answers "yes," provided that there are enough steps. Thus, the eye need not have developed in one or several jumps but rather could have developed in thousands or millions of small, nearly infinitesimal steps.

To visualize this process, work forward from $X^{(n)}$ to $X^{(n-1)}$ to $X^{(n-2)}$, and so on, to $X^{(3)}$ to X'' to X' to X to the eye. We can see evidence of the process in nature: Some single-celled animals have light-sensitive spots that give them a perception of light and darkness. Perhaps this light-sensitive spot enables them to detect the appearance of a predator and take evasive action. In certain many-celled animals, the light-sensitive area is recessed slightly, so the rudimentary eye has better directionality than a simple light-sensitive spot; this directionality conveys more information as to the whereabouts of the predator. A more deeply recessed eye gives more information about direction, so animals that have recessed eyes survive better than those with flatter eyes. Thus, the evolutionary pressure is toward recessed eyes.

As the animal evolves, the recess deepens and eventually almost closes; that is, the eye is now a hollow sphere with a light-sensitive interior and a small opening to the outside: a pinhole camera. Since these are underwater animals, the eye is filled with water. Perhaps the pinhole is covered by a protective membrane, and the material inside the eye develops a higher density than water; this gives the eye better focusing ability than a simple pinhole camera. In like manner, the sack that contains the fluid inside the eye detaches from the retina and develops into a lens. Ultimately, we get a full-fledged eye—what Dawkins calls a "camera eye" because it has a lens, an aperture, and a light-sensitive layer—as opposed to the rudimentary earlier eyes.

These intermediate stages are not figments of my imagination, nor of Dawkins's: Each intermediate stage I have described exists in molluscs. In addition, Brian Goodwin, [1994] a biologist at the Open University in the UK, shows that the eye develops in the embryo precisely similarly to the way in which the younger embryo develops its nervous system. That is, evolution did not have to invent a wholly new process to invent the eye but rather used a process that already existed. Goodwin calls such adaptations *morphogenesis* and shows how new forms are commonly derived from existing forms and almost never from scratch. Goodwin notes that similar but not identical eyes have developed independently more than 40 times. They are similar because they depend on already existing developmental processes. The mammalian eye, incidentally, evolved so that the

61

nerve cells lie on top of the retina, and the light has to pass through a layer of nerve cells before it strikes the retina. This is a little like putting the electrical wires on top of the light-sensitive screen in a television camera; no sensible designer would have done it that way.

Dawkins easily gives the lie to the creationists' claim that the eye must be fully formed and perfect or else it cannot work at all. Each intermediate stage "works" in some manner, even those that cannot form an image. Indeed, a great many human eyes (including mine!) are far from perfect: Otherwise, we would not need eyeglasses. Some eyes are so far from perfect that their acuity cannot be brought to normal even with corrective lenses. My elderly aunt, for example, suffered from macular degeneration. As a result, her visual acuity was more than 10 times poorer than normal, and she could neither read nor find her way around in an unfamiliar setting. She could nevertheless cook, turn lights on, use the telephone, dress herself, and generally live alone unaided. Her eyes, as imperfect as they were, were very useful to her, and she firmly believed that she was better off than if she had been wholly blind. Indeed, if you are a small mollusc, then one percent of an eye (a light sensitive spot) may be extremely useful to you if it signals the approach of a predator. So much for the argument that an eye has to be perfect or it is useless. To the contrary, the eye is the outstanding example of an organ that can be useful in the most rudimentary form imaginable.

* * *

The question, "What is half an eye?" is based on the assumption that biological systems are so complex that they are unlikely to develop from scratch. Thus, creationists often note that, if they wanted to build a Boeing 747, they would not throw all the parts into the air and hope that they came down as a fully assembled airplane. [Kelemen, 1990] You can calculate the odds against that, just as you can calculate the odds against building a bacterium by throwing all the chemicals into a pot and waiting. The odds are very slim indeed. So are the odds that, if you froze a droplet of water, it would solidify into a hexagonal structure. Yet that is precisely what happens when snowflakes form in the atmosphere.

As we see from Dawkins's exposition, however, no evolutionist claims that life developed in that way. Indeed, the 747 did not develop that way either. Before we had the 747, we had hot air balloons, then a series of disastrous failures to build heavier-than-air machines, and finally the Wright brothers' successful flight. Next, we developed biplanes, single-winged planes, warplanes, the Douglas DC-3, the jet fighter plane, the Boeing 707, and, finally, the 747. In a very real sense, it took over 100 years and countless intermediate steps to develop the Boeing 747. We no more built the 747 from scratch than nature built the eye from scratch. Similarly, we did not intend to develop the Boeing 747 any more than nature intended to develop either the human or the eye. These are merely the latest stages of developmental processes that could have taken any of a number of possible other turns. For example, if petroleum had not been so cheap, the state of the art in air travel today might be not a Boeing 747 but a dirigible, and we would take our long-distance trips on a very efficient rail system.

In another book, *River out of Eden*, [1995] Dawkins asks how many generations are necessary for a substantial evolutionary change. In other words, has there been enough time for the eye to develop roughly as I have described? Creationists presume not. In response, Dawkins describes a computer model by the two Swedish scientists, Dan Nilsson and Susanne Pelger.

Nilsson and Pelger [1994] call their paper, "A *Pessimistic* Estimate of the Time Required for an Eye to Evolve" [my italics]. They describe a computer model that begins with a flat, light-sensitive patch. The patch consists of a light-sensitive layer underneath a flat, transparent layer of protective tissue. Such a rudimentary eye is useful to an animal, even if it can do no better than allow the animal to detect the presence of a predator. If,

63

however, the eye were recessed, it would gain in directionality and thus aid in locating the direction of the predator. Hence, Nilsson and Pelger gradually varied the depth of the eye and then the diameter of the aperture and calculated the visual acuity (directionality) of the resulting eyes. When focusing caused a greater improvement than decreasing the aperture diameter, Nilsson and Pelger gradually changed the shape of the transparent membrane (the cornea) so that it would focus light onto the light-sensitive layer. Similarly, they allowed a gradual change in the index of refraction (light-bending ability) until a lens developed in the medium behind the cornea. Dawkins uses their model to elucidate the question whether there has been enough time for an eye to develop.

Nilsson and Pelger used a change of 1 percent per generation in their model. Dawkins describes the results of their program as "swift and decisive." Acuity increases steadily as the eye transforms itself from a light-sensitive patch to a deep cup to an eye with a cornea to an eye with an internal lens. Furthermore, the lens displays a varying index of refraction, precisely like lenses in living creatures, but unlike common glass lenses. Additionally, the intermediate eyes are not fictions; rather, each intermediate stage is known in the animal world.

To estimate how long it may have taken for an eye to develop, Nilsson and Pelger made certain assumptions about heritability. They assumed that only half of any change was heritable from one generation to the next. Dawkins considers this a conservative value in the sense that it exaggerates the time necessary for the eye to develop. Similarly, Nilsson and Pelger prohibited more than one change at a time, and they chose low numbers to describe both the frequency with which variations appear in the population and the survival advantage gained by improved vision.

Even so, the camera eye developed in less than 400,000 generations. Among many animals, a generation may be a year or less, so half a million years should be ample time for evolution to develop an eye with a cornea, a lens, and a retina.

Eyes first appear in the fossil record approximately 550 million years ago; Nilsson and Pelger note that eyes could have evolved more than 1500 times during that time. Thus, we may firmly refute the claim that the camera eye could not have developed within the available time.

* * *

We cannot trace the development of the eye through the fossil record because the eye is composed entirely of soft tissue, not bone, and soft tissue is rarely preserved. Thus, we have no empirical evidence that the eye in fact developed as evolutionists assume. So let us ask an easier question, What is half an ear? The intermediate steps are preserved in the fossil record.

Stephen Jay Gould is a paleontologist, a historian of science, and an essayist. He teaches at Harvard, is Honorary Curator in Invertebrates at the American Museum of Natural History, and writes a monthly column for *Natural History* magazine. In his article, "An Earful of Jaw," Gould [1993] explains how a bone that originated as a support for the gills of fishes gradually evolved into a brace that held the jaw to the skull. Eventually, in mammals, this bone became the stirrup in the middle ear. Along with other bones, the hammer and the anvil, the stirrup concentrates sounds detected by the eardrum and directs the sounds to the nervous system.

The details of the transformation of these bones from jaw bones to ear bones are too complex to describe in a few words. Let it suffice to note that Gould provides two bits of direct evidence for this transformation. First, he points out that the transition is documented, with intermediate stages, in the fossil record.

Second, the transition is reproduced in the embryonic development of mammals. That is, the ear in a mammalian embryo develops by passing through all of the earlier stages, a fact that had been noted before the development of evolutionary theory. For example, the mammalian embryo passes through a stage during which it develops gill arches; these gill arches then develop into the ear. This is considered evidence that the ear evolved through these stages and was not created out of whole cloth: If all the animals had been carefully designed or created at once, then you would expect the ear to develop straightforwardly and efficiently in the embryo. Instead, its development is circuitous and inefficient. Indeed, the human embryo passes through stages where it resembles a fish embryo, then a reptile embryo, then a pig embryo, before it finally begins to look human.

If the animals were not created at once, and all you had to work with was a gill arch or a reptilian jaw, then you might have made the mammalian ear by first building the gill arch, then the reptilian jaw, and then the ear. In short, if the mammalian ear had not developed from the gill arch, it is very unlikely that a mammalian embryo would pass through a stage where it first developed the gill arch. That process is too inefficient to suggest anything but trial and error. Thus, the existence of intermediate forms in embryonic development is strong evidence for evolution.

In addition, Gould notes that snakes have no eardrums. They hear primarily by putting their heads to the ground ("primarily" because they also hear through vibrations transmitted by their lungs). Vibrations are thus transmitted to the brain through the same bones that eventually became the mammalian hammer and stirrup.

Why did these bones move from the jaw of a reptile to the ear of a mammal? Timothy Rowe of the University of Austin has suggested that the move was related to the increased size of the skull due to the development of bigger brains. [Fischman, 1995] Rowe studied the development of certain reptilian ancestors of the mammals and concluded that the shifting of the ear bones coincided with the growth of the skull. As the skull expanded, the ear bones were carried away from the jaw. No longer needed to hinge the jaw and already useful in hearing, they were gradually optimized for hearing.

This section is headed "Taste," and I want to make the point that the validity of a scientific theory is not a matter of taste. A theory may be true no matter how distasteful its outcome; alternatively, a theory may be false no matter how much we would like to think otherwise. The same is true of a religious belief; it may be true even if we passionately doubt it, or it may be false even if we passionately believe it. The only statements we can make with near certainty are those derived from careful study of objective evidence.

Is religion falsifiable?

The philosopher Antony Flew, now of Reading University, recounts a parable about two people who chance upon a clearing in the forest. [Hick, 1964; Pojman, 1987] Both flowers and weeds grow in the clearing. One of the people, the Believer, says that some gardener must be tending the plot, whereas the Skeptic disagrees. They set up camp and watch, but no gardener appears. The Believer suggests that the gardener is invisible, so they patrol with bloodhounds, then set up an electric fence, but there is still no evidence of a gardener. The Believer insists, however, that there must be a gardener, even if that gardener is invisible, silent, odorless, and impervious to electric shocks. The Skeptic asks how that differs from an imaginary gardener or no gardener at all.

Flew uses this parable as a jumping-off point to discuss whether religion is falsifiable. Specifically, referring to the problems of evil and suffering, he asks what would have to happen to falsify a belief in God or in God's love. Flew's question is rhetorical; he clearly implies that nothing will falsify a firm religious belief. An Oxford philosopher, Basil Mitchell, agrees or, more accurately, admits that nothing can count decisively against the belief of the true believer; by definition, the believer is committed to a belief in God and is not a detached observer. [Pojman, 1987] That is, to Mitchell, the concept of falsifiability is not appropriately applied to a religious belief,

67

whereas, to Flew, its lack of falsifiability evidently counts against religion.

Another Oxford philosopher, R. M. Hare, responded to Flew with a parable of his own: A lunatic (Hare's word) believes that the dons want to kill him. A friend believes otherwise and tries to convince the lunatic by introducing him to the dons and showing him that they are friendly, gentle people and mean him ho harm. The lunatic responds that the dons are duplicitous and are really plotting against him, all the while pretending to be friendly.

Hare calls the lunatic's belief a *blik*. This is a term that Hare has coined to describe a belief that is neither verifiable nor falsifiable. Hare notes that the friend also has a blik: The friend's blik is that the dons are *not* planning to kill the lunatic. Hare considers this belief a blik just as much as the lunatic's belief is a blik. That is, the friend does not have no blik at all, but rather has the blik that the dons are harmless. Precisely like the lunatic, the friend cannot prove his blik, because the lunatic can always find an *ad hoc* hypothesis to refute the friend's arguments.

Hare's article was influential, but it seems to me that it contains within it the seeds of its own destruction. First, the issue is not whether a sane person can convince a lunatic that the lunatic's blik is wrong; he cannot. The issue is, rather, what arguments could both the friend and the lunatic use to convince a detached observer which one is right. In this case, it is clear that the detached observer would rule in favor of the friend, not the lunatic, because the friend would present more convincing evidence.

Later in the article, Hare notes his own blik that the steering column of his car will not fail when he goes for a drive. This blik gives him confidence, without which he might be paralyzed into inaction. Hare's confidence might be based on a blik, but I have no such blik. Whenever I drive my car, I am perfectly aware that the steering column might fail. I am equally aware, however, that the vast majority of steering columns do not fail during normal use, so I drive my car in the uncertain knowledge that the steering column will probably not fail. This is not a blik; it is a statistical statement based on evidence, which I see all around me, that other cars have sound steering columns. Not all firmly held beliefs are bliks.

Hare's position is that a religious belief need not be defended because it is a blik and can neither be proved nor disproved. Hare himself, however, distinguishes between bliks that are right and bliks that are wrong. Indeed, he seems to intend his lunatic to be analogous to the religious believer who supports his belief with *ad hoc* hypotheses. The issue, then, is not whether people have bliks but rather whether their bliks are right or wrong. How do we determine whether bliks are right or wrong? We look for evidence. Far from refuting Flew's argument, Hare has strengthened it.

* * *

Ian Barbour, Professor of Science, Technology, and Society at Carleton College and formerly Visiting Scholar in philosophy of science and in divinity at King's College, Cambridge, takes a position that is in some sense midway between those of Flew and Mitchell. In his book, *Myths, Models and Paradigms*, [1974] Barbour notes that even scientific theories are not truly falsifiable. No crucial experiment can distinguish with certainty between rival theories, in part because their proponents can always make up an *ad hoc* hypothesis. In addition, all theories have exceptions or anomalies, as I noted above, so theories or paradigms are never proved conclusively. Rather, they are accepted because they can explain a preponderance of the available evidence. Thus, Flew is being perhaps a bit too stringent in his demand for falsifiability. Barbour agrees, however, that empirical evidence is not irrelevant to religious belief.

Following Thomas Kuhn, Barbour notes that established theories (as opposed to fledgling theories) are not rejected by

69

being falsified but rather are rejected only if they can be replaced by more promising or more complete alternatives. Thus, he argues that an established religious paradigm cannot be simply falsified but would instead have to be replaced by a competing paradigm, a paradigm that explained the available evidence better. There are, however, no rules for choosing between paradigms; rather, individual judgement is necessary. Since there are no rules for judging, there can sometimes be disagreement even among people who have access to the same data and the same theoretical constructions.

Barbour argues, therefore, that scientific beliefs are subjective in a certain sense. (1) The paradigm of the observer determines in part what that observer sees. When Aristotle saw a pendulum swinging, he saw an object slowly attaining its natural state of rest, whereas Galileo saw an object whose inertia would keep it swinging forever if it were not for the force of friction. (2) A paradigm's resistance to falsification is another subjective aspect, as is, finally, (3) the lack of rules for choosing between paradigms.

Barbour argues further that, just as science has subjective aspects, religion has objective aspects: (4) common data on which we all agree, such as the existence of fossils, (5) the use of evidence to support a contention, as when a person is "miraculously" saved from an accident, and (6) criteria that fall outside all paradigms, such as the demand for a simple and coherent explanation.

To Barbour, religion and science share all six of these aspects, but religious beliefs are more subjective [aspects (1) through (3)] and less objective [aspects (4) through (6)]. Both scientific paradigms and religious paradigms are highly resistant to falsification or replacement, for example, but religious paradigms are much more highly resistant. To put it another way, scientists are more likely to give up or replace one of their beliefs than are, say, clergypersons.

On the other hand, religion lacks a base of primitive or underlying laws, like Newton's laws of motion or Maxwell's equations of electromagnetism, which are not in dispute. Such laws are in close agreement with observation and could in principle be falsified by a mere handful of counterexamples; they are as close as we come to "scientific truth." There is no "religious truth" in the sense of a set of basic principles to which all agree. This seems to me to be a major and important difference between scientific and religious beliefs.

Nevertheless, I agree in principle with Barbour's analysis, though I think he may minimize the subjective nature of at least some religious beliefs and exaggerate their objective nature. Earlier, when I outlined Popper's theory of falsifiability, I noted Popper's comment that some genuinely falsifiable scientific theories are nevertheless propped up by their "admirers," who use various stratagems and therefore lose their claim to objectivity. In religion, where there is substantially less hard evidence than there is in science and no primitive laws or facts, it is very easy for the "admirers" of a specific religious belief to employ *ad hoc* hypotheses and other stratagems to defend their beliefs against even a mountain of evidence. I gave an example in the preceding section, "Taste"; the evidence in favor of evolution is overwhelming, yet some reject it on the basis of specious arguments motivated purely by religious beliefs. In this sense, Barbour is right in principle, but Flew is right in practice: Too many people have set up their religious beliefs to make them unfalsifiable and will not change their minds no matter what the evidence.

* * *

In this chapter, I have shown that a scientific method is the only reliable way of distinguishing between facts and unsupported statements. Science is not perfect, but it has a greater claim to reliability than other belief systems because it is constantly questioning its own conclusions. Those who do not question their conclusions end up, like the homeopaths, clinging

71

to a belief so absurd that they have to invoke major but unspecified flaws in firmly established theories.

In the next four chapters, I will apply scientific method to the claims of religion and philosophy of religion. First, I will examine the quality of the evidence that lay people and the clergy bring to bear on the existence of God. Then I will examine the Bible for evidence that it is historically accurate, let alone the literal word of God. In a separate chapter, I will consider the problem of evil and show that the existence of evil is not a problem for the scientist. Finally, I will examine the Arguments of the philosophers and show that none is convincing in light of modern scientific theories.

Let us see how well these arguments (and Arguments) hold up when they are subjected to the kind of scrutiny that scientists expect of each other.

God said it
I believe it
That settles it

BUMPER STICKER

Don't believe everything you THINK
ANOTHER BUMPER STICKER

Unless we believe what they cannot prove, we are not
good people.
JOHN MURPHY

Extraordinary claims require extraordinary proof.
MARCELLO TRUZZI

Chapter 3
Signs, wonders, and anecdotes

Here I will distinguish between *popular* religious beliefs,
which may be held by laypersons, and those espoused by
theologians and philosophers, though the dividing line is often
blurry. I mean nothing pejorative by the term "popular," but
rather mean to classify certain beliefs that may be held by people
who, for example, have no formal training and have never heard
of the Argument from First Cause. Indeed, the statistical
argument that I will describe shortly is very sophisticated, and I
include it among popular beliefs only because it complements a
discussion of signs from God.

Miracles

I saw the following cartoon in *Physics Today* [November,
1994]: A fortune teller or an astrologer tells a client, "You have
little understanding of probability, causation and coincidence."
She does not have enough clients.

A man from Denver was in a bank across the street from the Federal building in Oklahoma City when a bomb detonated. He was a Red Cross volunteer and was trained in cardiopulmonary resuscitation and first aid, so he spent the next two days working with rescue crews. He told the press that *one reason* he had escaped injury was to help those who were hurt. I doubt that he knew of another volunteer, a nurse, who left her home, her four children, and her husband, and rushed to the scene only to be killed by falling debris. [Cohen, 1995] The 37-year-old nurse had only recently graduated from nursing school and remarried. The press report conveys the image of a person who was dedicated to the welfare of others and once quit a debt-collecting job because of the pain she caused her clients. If we follow the logic of the Denver man, we must conclude that *one reason* that this apparently worthy woman had escaped injury and come to the scene of the blast was so that she could later be killed by falling debris. Or are we to believe that the Denver man was somehow worthy, whereas the nurse was not?

The University of Colorado football team defeated Michigan on a "miraculous" catch by Michael Westbrook. The Boulder *Daily Camera* asked Westbrook if he believed in miracles. "Yes, I do," he replied. "Who could've asked for more?" Another player, Christian Fauria, said, "I don't want to get religious, but we've really got to thank God. He had a lot to do with it. I don't know whether it was divine intervention or what." A month later, the same team was humiliated by Nebraska. Fauria said, "We played horrible [sic] and they played great."

God's interest in football extends even to the high school level, if we are to believe Phil Bravo, the coach of the Centaurus (Colorado) Warriors. When his team defeated the defending state champions in the quarter-final of the state tournament, Bravo said, "Today's game was nothing short of miraculous. All I can do is give thanks and glory to God." He did not mention God when his team lost in the finals.

Do Fauria, Westbrook, and Bravo really think that God influences the outcome of football games and that the teams that win are always those whose players or coaches are the most deserving?

Michael Chang won the 1989 French Open tennis championship when he was 17. His semi-final match against Ivan Lendl was bizarre in that he apparently "psyched" Lendl by serving underhand and then by crowding the service line and forcing Lendl to double-fault at the end of the match. In 1995, Chang had this to say:

> ...It was more of the man upstairs than anything else.... I always look back on '89, and [I say] that '89 was something that was meant to be. I think that at that time, during the situation at Tiananmen Square—they had the massacre there and it was a very down time for the Chinese people, not just in China but around the world—and I think that it was something that the Lord wanted to happen so that people could take their mind[s] off that [the massacre] for a little while and see that there is something good happening to a Chinese person halfway around the world.

John McEnroe said (tongue in cheek, I think) that Chang was maybe "onto" something, since he had never won another Grand Slam tournament (one of the four most important annual tennis tournaments); as of this writing, he still has not. For my part, I wonder how Chang could compare a mere tennis tournament with a massacre and act as if his victory had been a suitable consolation prize. Why didn't God simply call off the massacre?

A clergyman friend of mine works in a hospital. He told me an anecdote about a woman, a trained nurse, whose baby was dying. The nurse heard the wings of the baby's guardian angel flapping frantically; she rushed into the baby's room in time to save his life.

75

Finally, a press report about Dan Wakefield's book *Expect a Miracle* tells of a "miracle" that happened to Michael Crichton, the novelist and movie director (and a trained physician). Apparently, Crichton had a "feeling that something wonderful" was going to happen to him. He was "struck dumb" when a certain actress walked in; they eventually married. Wakefield describes the case of a woman with multiple sclerosis; she was taken to the British equivalent of Lourdes and walked away free of the disease. Wakefield himself calls his recovery from alcoholism a miracle.

Are these miracles? Maybe, depending what you mean by "miracle." I think we can rule out Crichton's experience out of hand. We all probably have "premonitions" all the time; most of them are fantasies that do not correspond with reality and are soon forgotten. We remember selectively and sometimes incorrectly, and we attribute significance to our memories, whether that attribution is deserved or not.

Consider, for example, the Senior Prom at Boulder High School. Let us suppose that 300 students attend it each year. They have approximately 600 parents. Perhaps half of the parents go to bed at eleven or so but can't sleep. They are convinced that the next person they meet will be a police officer who informs them that their child has been killed in a fiery auto crash. Nothing happens for 10 years; then one student is killed in a fiery auto crash. That father reports to the newspaper that he had a premonition, and everyone agrees that he foresaw the event.

Did he? I can't tell you for certain that he did not, but I *can* tell you that, in our example, 2999 parents over a ten-year period had similar, but false, premonitions. It certainly seems most likely that the "premonition" was merely a coincidence and that the father and the newspaper reporter have "little understanding of probability, causation and coincidence."

More pertinently to Crichton's anecdote, memory is apparently very malleable. It is easy for an experimenter to implant memories in a subject, for example. [Loftus, 1995]

76

Minouche Kandel, a lawyer, and Eric Kandel of Columbia University (and a 2000 Nobel Laureate in physiology or medicine) present the case that repressed and recovered memories may be real. [Kandel and Kandel, 1994] They nevertheless describe the case of Paul Ingram, who was arrested after his daughters accused him of Satanic abuse. Ingram initially did not remember committing the alleged crimes, but he was told that sometimes offenders repress memories of abuse. Eventually, he produced "memories" and confessed to the crimes.

Richard Ofshe, a psychologist hired by the prosecution, was evidently suspicious. Ofshe concocted a story that Ingram had forced his son and daughter to commit incest, and told the story to Ingram. At first, Ingram denied the story, but Ofshe told Ingram to imagine that the story had happened and to "pray on it" as he had prayed on other stories that led to his confession. Shortly afterward, Ingram reported detailed "memories" of the fictitious story. Ingram's son and daughter denied the story, but Ingram, who was apparently highly suggestible, denied that he had been influenced by Ofshe. Ingram is now in jail for committing crimes that almost certainly never happened. Other people have been told false stories about getting lost in a shopping mall or injuring themselves; soon, they too had detailed "memories" of these events, even though the events had never happened.

Finally, Bruce Bower [1996] cites research on how our memories distort or elaborate on actual events. A group of college students was given a list of related words. Later they were given a closely related word that was not on the original list but often incorrectly "remembered" the word anyway. Bower also reports studies of memories that are strengthened or elaborated by hindsight. Suppose, for example, you meet a new associate and form a good initial impression. The associate later becomes unusually successful, and you remember not merely having formed a good impression but, incorrectly, that you had

77

thought of the associate as a rising star. In short, memories can be false, or they can be embellished by hindsight.

Even if Crichton really remembers accurately having the feeling *before* the actress walked in, his anecdote is weak. When I went to my friend's wedding, if memory is an accurate guide, I expected absolutely nothing. I met a woman there and within several days decided I would marry her, provided that she was of the same opinion. We have been married for over thirty years. Did a miracle happen to Crichton but not to me? If Crichton divorces, do we retrospectively declare the miracle void?

The nurse's testimony is at least equally weak. Even if the baby really was dying, a claim for which we have only her statement, we have no evidence that the baby's guardian angel intervened. Perhaps the baby himself rustled his blankets or coughed or gagged, and the nurse later misinterpreted the sounds as the sounds of an angel flapping its wings. If you think that is impossible, go back to the Müller-Lyer illusion; your ears can be deceived as easily as your eyes.

Similarly, I see no objective evidence for ascribing Wakefield's cure to a miracle. Indeed, doing so takes credit from Wakefield himself, who very probably had already decided to give up alcohol and used his church and his friends there as an aid. Consider, instead, a report by Diana Nyad [1995] on National Public Radio. Nyad, a former long-distance swimmer, told of a disabled man who prayed that his mother would survive cancer, because no one else had ever hugged him. His prayers were in vain. His mother died, and he got depressed and tried to kill himself. Afterward, he took a close look at himself and realized that the problem was him, not his disability. He was simply not fun to be with, so he changed his attitude and evidently began to find friends, because he is now married and, not incidentally, gets hugged by his wife. This man claims no miracle, but rather believes that he himself analyzed his situation and changed it. There is no evidence for a miracle and no logical reason to ascribe the man's change to anything outside himself.

There is similarly no reason to ascribe Wakefield's cure to anything external to Wakefield and his circle of friends.

People who are cured of debilitating diseases are another matter. Some may truly be cured; many diseases have spontaneous remissions. That we do not understand them is no evidence of a miracle, at least not in the sense of divine intervention. In addition, in these cases, sometimes what is lacking is hard evidence that the patient really had the disease and that it was diagnosed by a physician. Martin Gardner, [1989] the author of many books and the former author of the Mathematical Games column in *Scientific American*, examined a claim by the faith healer and evangelist Oral Roberts. Roberts says that he contracted tuberculosis and began bleeding from his nostrils. Later on, he went to a faith healer, who touched his head. Roberts leaped up and shouted, "I am healed!" Unfortunately, Gardner has been able to find no evidence that Roberts had ever contracted tuberculosis. Gardner says that he can find only statements that Roberts did *not* have tuberculosis, but these were recorded only after his alleged cure. There is apparently no record other than Roberts's own statements that he has ever had tuberculosis.

According to an anecdote I have read (and which may be apocryphal), the novelist and skeptic Anatole France once visited the famous shrine at Lourdes. He saw the room where they kept crutches that had been discarded by those who were supposedly cured there. Where, France asked, do they keep the wooden legs? In other words, where is the evidence of a *real* miracle, a miracle that is unambiguous and is not susceptible to a naturalistic explanation?

I have discussed anecdotal evidence earlier (see "Anecdotal evidence," Chapter 2). I argued that anecdotal evidence can never be conclusive, that it should give you a hunch as to what to study. Further, anecdotal evidence can be tainted by selectively citing only confirmatory anecdotes. Anecdotes that are based on memory can similarly be tainted by selective or incorrect recollections. For this reason, I cannot accept as serious evidence anecdotes that may well describe mere coincidences (if they are true at all).

Credit and blame

We have all heard people sigh, "It must have been God's will," after a tragedy of some sort. If it gives comfort, that view has value. But can it stand scrutiny after the pain of the tragedy has lessened?

On Palm Sunday, 1994, a series of tornados killed 44 people and injured 350 in Alabama, Georgia, and the Carolinas. In particular, a tornado destroyed the Goshen Methodist Church during religious services, killed 20, and injured 80. Six of the 20 dead were children. [Bragg, 1994] One of the congregants asked the *New York Times*, "We are trained from birth not to question God. But why? Why a church? Why those little children? Why? Why? Why?" "If that don't shake your faith, nothing will," said another congregant.

Apparently, nothing will. A third congregant noted that a friend of his rarely went to church and had not planned to go that Sunday but later changed his mind and went. The friend, his wife, and their daughter were all killed when the church collapsed. "Maybe that's what people mean when they say God works in mysterious ways. I know the [man]. He could not have lived if his wife and child were gone." This congregant gives God credit for "allowing" the man to die with his family but does not blame God for allowing the tornado to destroy the church nor for allowing the family to be inside the church during the disaster. If God was not able to stop the destruction, why did he not keep the entire family away from the church, instead of sending the father to his death? Another congregant implicitly answered: "You see, God took them because he knew they were ready to go. He's just giving all the rest of us a second chance." Does he really believe that six children were "ready to go"? If they were, then why were their deaths a tragedy?

These congregants believe firmly in God's goodness, and they will invent any manner of *ad hoc* hypothesis in order to sustain their belief. As Popper argues, they lose claim to objectivity when their claims become untestable. In this case, God is compassionate, not because he allowed a family to be killed, but because he *sent* a man to be killed. What would God have done if God had been evil? Either hypothesis—that God is compassionate and that God is evil—yields the same result. The belief is therefore unfalsifiable. (In fairness, the ministers, a married couple who lost their daughter, were far more sophisticated than their congregants and distinguished between God's laws and the laws of nature. To them, God could not control the tornado. Their God is a "God of hope" and never wills anyone to die. For a discussion of this view, see "Reward and punishment," below.)

Samuel Proctor is a Baptist minister and Professor at Duke and Vanderbilt Divinity Schools and at Rutgers University. He appeared as part of a panel discussion on the Bill Moyers series *Genesis* [1996] on the Public Broadcasting System. The panel of clergypersons and writers was discussing Noah and the Biblical Flood. The discussion was wide-ranging, but at least some of the panelists thought that God had been too hard on humanity; in their view, not all the people could have been evil enough to warrant such wholesale destruction. Some argued further that Noah could not have been as righteous as the Bible claims, since he did not once question God or defend the other people, as had Abraham when God wanted to destroy Sodom and Gomorrah. Not so Proctor: Alone among the panelists, Proctor argued that the real story of the Flood was not that God destroyed the world, but rather that God gave humanity a second chance. To Proctor, God is perfection and cannot be criticized, even for a deed that other clergypersons saw as a monstrous act on the scale of the Holocaust. Some of the panelists considered the episode a puzzle, even if it did not shake their faith. Proctor, however, gives God the credit for saving a half-dozen people but not the blame for destroying countless others. Humanity may have got a second chance, but what about the majority of individual human beings who did not? Proctor is guilty of selective use of the evidence: He looks at the few who were saved, defines them as "humanity," and ignores the thousands who were destroyed.

Proctor and the congregants of Goshen Methodist Church give God credit for his good deeds but do not blame him for his bad deeds. Instead, they either ignore God's bad deeds (Proctor) or explain them away (the third congregant). Proctor and the congregants use or interpret the evidence selectively and make their hypotheses unfalsifiable. Their beliefs cannot stand scrutiny.

Reward and punishment

Why righteous people suffer is a serious problem for those who believe in miracles and in a personal god who protects them. Harold Kushner deals with this problem in his book *When Bad Things Happen to Good People* [1981]. A best seller, this book reportedly performed a great service to a great many people who would otherwise have blamed themselves for their misfortunes. Kushner told us (correctly, I think) that things happen randomly or for reasons that we cannot predict. If misfortune strikes, it strikes; it is not your fault; misfortunes hit good people as well as bad.

Kushner is a rabbi and no stranger to misfortune. In his book, he gives many examples of people who assumed or were told, sometimes explicitly, that they must in some way be unworthy if they are suffering. Others believe or are told that their suffering must be for a purpose. Kushner gives the example of a young woman he calls Helen, who has been diagnosed with multiple sclerosis. Helen, not unreasonably, asks, "Why me?" She has tried to be a good person; she has children and a husband who need her. Why her? By way of consolation, her husband tells her that God must have his reasons. Helen takes no consolation in this thought; to the contrary, she feels guilty for being angry at God. How can she pray to God to cure her of her disease, when it is God who gave her the disease in the first place? Helen ends up hating God.

A belief in divine reward and punishment or in some underlying purpose to your suffering can be devastating. That is not a valid reason to reject such a belief; as I have pointed out earlier, a proposition is not necessarily false because you do not like it. Nevertheless, Kushner rejects these beliefs because of the harm they do to people. And I have to agree with him, in part because there is no evidence to support such beliefs.

Kushner was a young rabbi who, by his own admission, had never seriously evaluated his own beliefs until a terrible personal tragedy caused him to rethink his entire worldview: His baby son Aaron was diagnosed with a rare disease called *progeria*, or rapid aging. Aaron grew to be about 3 feet tall, he was bald, and he grew old and died within a decade or so of his diagnosis. He suffered from angina and eventually died of a heart attack, but his emotional pain, according to Kushner, was at least equal to his physical pain.

Predictably, some people suggested that God might have had a reason for giving Aaron this disease: Perhaps it made Kushner a better, more compassionate rabbi. For example, they argued, think of all the people on whom Kushner has had a positive influence as a direct result of Aaron's illness.

My mother had Alzheimer's disease from 1976, when she was 67, until her death in 1998. In the late 1980's, after my father had retired from his job in order to take care of her, someone suggested that perhaps God had given his wife Alzheimer's disease in order to test him. My father's answer, "I wish God would just test *me* then and leave my poor wife alone," was more measured than mine would have been. I was outraged at the suggestion and can only guess at the outrage Kushner must have felt at the suggestion that a mere child was given a dreadful disease specifically to make Kushner a better rabbi. Had God no better, more compassionate ways to make Kushner a good rabbi?

Yet Richard Swinburne, [1996] Professor of Philosophy at the University of Oxford, argues substantially as Kushner's and

83

my father's tormentors: that God has created the best of all *possible* worlds and that sometimes suffering may be a good thing if it leads to something positive. Obviously this is true at some level; if I break your arm while rescuing you from a speeding train, you will probably be grateful for the suffering. If you were a child, however, I might not be able to explain why I had broken your arm, and you might resent me for it. Theists argue analogously that suffering must occur for a reason that we do not know or cannot understand but which is valid nevertheless.

Swinburne admits that someone's suffering may well be for the common good, not for the sufferer's good. He argues that, without suffering and evil, we could not exercise our free will, at least in any meaningful fashion. It is knowledge of suffering and evil that enables us to make significant choices—or, rather, in the absence of evil or potential evil, our choices are relatively insignificant. He leaves unexplained why God could not have created us with just a little less tendency to evil.

Swinburne makes the case for a connection between human evil and free will. But he does not convincingly address the connection between random misfortune and free will. Lou Gehrig, for example, was struck down at the height of his career by a disease that has become eponymous with him; so was the pitcher Catfish Hunter. Jacqueline DuPré, the noted cellist and wife of the conductor Daniel Barenboim, was similarly struck down by multiple sclerosis. How do these tragedies relate to Gehrig's, Hunter's, or DuPré's free will? If they intended to continue their careers, then the diseases deprived them of that option, that is, deprived them of free will. But Swinburne tries to get God off the hook by claiming that it is logically impossible for God to know the future.

By logically impossible, Swinburne means that even an omnipotent God cannot do something that is wholly inconsistent, such as make $2 + 2 = 5$, make a circular square, or create an object so massive that an irresistible force cannot move it. Swinburne's God, however, pervades the universe and knows exactly the motions of and forces on every particle in the universe. If that is so and if God has created a deterministic universe, then it is logically impossible for God *not* to know the future. If, by hypothesis, humans have what Swinburne calls "limited free will," then God may not know the future exactly, but free will is, here, an *ad hoc* hypothesis. It is not *logically* impossible for God to know the future; it is *by hypothesis* not possible.

I do not, in any case, see how the argument is relevant. If I see a young woman in the early stages of multiple sclerosis, I do not have to be a prophet, let alone a god, to be able to predict her long-term future with considerable confidence. I do not have to know the future to see someone suffering and want to relieve her suffering; I surely do not have to know the future exactly. Swinburne's God, however, is powerless: omnipotent but hung up on a philosophical argument.

Kushner, by contrast, rejected the image of God as an omnipotent and omniscient being, and reevaluated his beliefs. His argument is something like this: Babies and children in Kushner's view could not possibly be sinners (though he is not explicit about this point, I think we may assume that Kushner does not believe in past lives or original sin; such hypotheses could mitigate the view that babies cannot be sinners). Babies and children nevertheless sometimes suffer what seem to be cruel and debilitating diseases. Even when we see an adult or an older child suffer, we cannot imagine a crime serious enough to make God punish him or her in that way. If one of your children displeased you, would you give him multiple sclerosis? If one of your colleagues displeased you, would you give her son leukemia? Then why do you assume that God, whom you perhaps consider compassionate, would do so?

We all know of examples where apparently worthy people suffer great misfortunes, whereas evil people prosper. Kushner considers disasters such as train wrecks and airplane crashes. He considers cases where, let us say, 399 people are killed in an airplane crash, but one is "miraculously" spared. Often, that person avers that God saved him or her for one reason or another. Kushner finds it extremely unlikely that all the victims were evil and somehow deserved to be killed or maimed. He cannot believe that one person was necessarily more deserving than all of the other victims and concludes that misfortunes of this type are no more than random occurrences, neither predictable nor determined by God.

Ultimately, Kushner rejects the idea of divine reward and punishment.

Kushner argues that, if God knowingly allows innocent people to suffer, then he cannot be both omnipotent and benevolent. Kushner phrases his argument in terms of Job's guilt or innocence, but his argument boils down to considering three possibilities:

1. *God is omnipotent.*
2. *God is omniscient.*
3. *God is benevolent.*

You may pick any two of those three. If God is both omnipotent and omniscient, then he must not be benevolent; there is too much evil in the world to allow that belief. Kushner says that he could fear such a God but not worship him. He therefore decides that God is benevolent but not omnipotent.

Note my use of the word *decides.* Kushner provides not a shred of evidence or even reasoned philosophical argument that God is benevolent nor, on the other hand, a shred of evidence that God is not omnipotent. He simply makes up his mind that that is the way it must be: Like Ricky, he seems to think that if he believes something, then it is true.

Kushner has replaced the omnipotent God who gets all the credit and none of the blame with a sort of charmingly incompetent God who feels great compassion for us but cannot do anything concrete for us. (For a more scholarly version of Kushner's arguments, see, for example, Peacocke [1993].) Kushner has done a great service by debunking the kind of religion that makes everything your fault, but he has provided no real argument in favor of his point of view either.

* * *

In 1985, I was traveling in France with my daughter, whom I will call Shelley. Shelley was about to turn 15 and played on the girls' volleyball team at her school. She sometimes complained of pain in her wrists and, sporadically, in her knees, but we associated these pains with overuse or tendinitis. When we were in France, however, it seemed as if all her joints swelled up at once: knees, elbows, feet, knuckles, even her collarbone. We took her to a hospital in Nice, and they told us to take her home immediately.

Shelley and I ended up flying standby to Chicago, where we had to take a hotel for the night. Shelley naturally asked Helen's question, "Why me?" "I don't know, Shelley," I said; "it's nothing to do with you. There just is no reason." It was, as you can imagine, not a very satisfying answer.

In the morning, after I had had time to ponder the question some more, I told her I had a new answer: "Why *not* you?" When I said that, we both realized that, just as there was no reason for her to be singled out for a disease, neither was there any reason for her not to contract a disease. At 15, Shelley was beginning to understand the universe better, in my opinion, than those who believe in miracles, reward and punishment, and divine intervention.

* * *

Kushner asks the question, "Where was God in Auschwitz?" and answers it in the way that has become conventional for the liberal clergy: God was with the survivors. God, for Kushner as for many other liberal clergypersons, is the "power" that gives us the strength to endure adversity, to strive to understand the universe, to fight against war and oppression, to develop new medicines, and so on.

Using "God" in this way seems to me to be almost an allegory for what is the best in human beings, but Kushner does not use the term allegorically. Indeed, he notes his belief that "the part of us which is not physical" *cannot* die. He provides no argument that any part of us is nonphysical, let alone that it cannot die. Before making the statement that it cannot die, you have to establish whether it exists. As I will show below, I find no evidence that we are anything but biological organisms; that is, I find no evidence for the claim that we have a nonphysical part, so I hypothesize that it does not exist. Indeed, I doubt that anything exists outside the physical world. Still, I can draw an analogy that shows how a nonphysical entity can die: A song is a nonphysical entity of a sort. Suppose that everyone forgets a certain song and also that every copy of it is ultimately destroyed. Then that song has died. This is only an analogy, but it illustrates how even a nonphysical entity can cease to exist. Thus, I see no reason to automatically assume that the nonphysical part of us, even if it exists, is immortal.

Kushner, at any rate, makes precisely the same mistake he debunks earlier: He considers that God gave the survivors the strength to resist and to survive in Auschwitz. Let us suppose that, out of a group of 400 prisoners, 399 perished but one resisted and survived. Why is it plausible that God was with that one and abandoned the others? How does this example differ from the airplane crash that had 399 victims and one survivor? It does not differ one whit and leads me to conclude that God had no more to do with survival in Auschwitz than he had to do with any other isolated examples of survival against the odds. There is simply no reason to believe that the good or the pious or the deserving survived more frequently than the evil or the impious or the undeserving.

The liberal clergy's idea of God seems to me as unfounded as the God of miracles and the God of reward and punishment that they reject. It relies on selective use of evidence and on unfounded hypotheses.

Wishful thinking

I remember once, when I was in college, sitting outside the dorm and arguing religion with my friend Alan. I do not remember exactly what we were saying nor even whether we were making much sense. As we were talking, another student I did not know came along, apparently as I was arguing that the universe was guided by impersonal laws. The third student immediately asked, "Is that the kind of universe you want? A universe that has no purpose to it?" My response was something like, "What kind of universe I want is not the issue. What matters is the kind of universe that exists."

The student thought that the universe must be guided by some sort of purposeful force because he did not like to think otherwise. He is by no means alone in this sort of reasoning. In the first chapter of their book *The Nine Questions People Ask about Judaism*, rabbis and authors Dennis Prager and Joseph Telushkin [1986] present essentially the same argument.

Under the heading, "The need to posit God's existence," Prager and Telushkin argue that, without God, there can be no morality beyond personal preference. That is, if there is no God, then there can be no universal code of morality. True enough, I think, but irrelevant. There can still be an agreed-upon code of morality; since we do not know the universal code, an agreed-upon code is the best we can do. Even those who assume that the Bible is the universal code of morality have to interpret its meaning, as is made plain by the fact that not all of them agree. Applying the universal code is another problem, in any case, because a great many moral choices are not black and white but rather involve subtle distinctions that depend on precise circumstances. Thus, the universal code, if it exists, will have to be infinitely long and therefore impossible to understand in a finite time, let alone to apply.

In addition, postulating a God to give us a moral code leaves us with serious philosophical questions. If God gave us the moral code, why did he choose the code he did? Is something moral because God says it is, or does God say something is moral because it is? That is, did God choose arbitrarily, or did he have no choice? If God chose arbitrarily, why did he choose as he did? If God had said that murder was ethical, would it then be ethical? On the other hand, if God had no choice, then why did he have no choice? What is it that makes something ethical and something else unethical? Assuming that the moral code was given by God raises unanswerable and unilluminating questions, and solves nothing.

Rather than a single, agreed-upon code of morality, we encounter a number of codes that are sometimes similar or overlapping and sometimes at variance with each other. The fact that apparently thoughtful and humane people differ on major points suggests to me that there is in fact no such thing as a universal code that is applicable in all cases and to all people.

Let us assume, for argument's sake, that we agree that abortion is morally wrong. In specific instances, however, we might countenance abortion in order to save the life of the mother, who may have uterine cancer or some other condition that will make childbearing fatal. That is, even people generally opposed to abortion may well make an exception in certain cases, knowing fully that the fetus will be killed. The abortion is allowed as long as there is an otherwise legitimate reason for the procedure. Similarly, during war it may become necessary to bomb a city in order to destroy a munitions plant, even though many civilians will be harmed. The bombing is allowed as long as it has a legitimate purpose other than killing civilians. These are examples of *situational ethics*: the decision to perform the abortion or to bomb the city depends on conditions at the moment the decision is made. It is hard to see how such decisions could be codified into a pre-existing or universal moral code that could be applied to all possible cases.

But that is not the point. The point is that Prager and Telushkin do not like situational ethics, and they wishfully posit a God to deliver them a universal code. They present not a shred of evidence in favor of the existence of a universal code, but rather show that pragmatism does not necessarily lead to ethical behavior: a straw man in that few people believe that it does.

Prager and Telushkin quote the atheist philosopher Bertrand Russell as conceding (their word), "I cannot see how to refute the arguments for the subjectivity of ethical values but I find myself incapable of believing that all that is wrong with wanton cruelty is that I don't like it." They use this quotation to illustrate their claim that atheism is inherently amoral, that is, that atheism has no strict moral code, not that it is *im*moral. Whereas they agree that there can be moral atheists, they claim that such morality is based on secular values, not on "three thousand years of religion-based morality," as if mere antiquity somehow conveyed authenticity.

In any case, arguing that religion automatically conveys morality puts them on very thin ice. From the slaying of the men of Shechem in Genesis 34, to the massacre of Muhammad's enemies in Medina, to the Crusades, to the St. Bartholomew's Day Massacre in 1572 (when thousands of French Protestants were murdered in their beds), to the recent war in Yugoslavia (a purely religious war among people who speak the same language, Serbo-Croatian, and are indistinguishable except for their religions), religion has been responsible for perhaps more "wanton cruelty" than any other single force in society. Indeed, the physicist and evangelical Christian writer Hugh Ross [1998] explicitly approves of the massacres described in Genesis: The people who were massacred by the Israelites or drowned in the Noachian Flood were reprobates, and Ross likens their genocidal murder to surgery for excising cancerous tissue from the body. [Young, 2001]

91

* * *

The journalist Patrick Glynn [1998] argues to the contrary that religion makes people more moral or humane. His best example is the contrast between the American and French Revolutions. Glynn argues that the founders of the American republic were religiously motivated, whereas the leaders of the French Revolution were atheists. In part for that reason, says Glynn, the French Revolution culminated in a bloodbath known as the Reign of Terror, whereas the American Revolution was comparatively free of atrocities.

Biblical literalists who claim Jefferson and Washington as theirs are greatly exaggerating. Nevertheless, Jefferson, Washington, and even Thomas Paine were deists who were motivated in part by a belief in *natural religion*, that is, the belief that religious knowledge is accessible to human reason. For example, when Jefferson wrote that all men were "endowed by their Creator with certain unalienable rights" and that God gave us "life and liberty at the same time," he was not being metaphorical.

Deism covered a broad range of beliefs, but the majority of deists did not believe simply that God had created the world and then disappeared. Rather, deists believed in a supreme being but worshipped him by their actions rather than by engaging in rituals. They believed in the moral teaching of the Bible but not its historical accuracy. They did not believe in revelation. They rejected much of the practice and symbolism of the established religions and regarded all religions as similar at their cores. Thus, there is no doubt that the founders of the American republic were religiously motivated.

The French Revolution was inspired in part by the authors of the *Encyclopédie*, a liberal and skeptical work that opposed the abuses of both the Church and the monarchy. The authors of the *Encyclopédie* were mostly deists, though the editor, Denis Diderot, in later life became an atheist. Thus, though French deism was perhaps more inclined toward atheism than English or

American, it is not fair to call the leaders of the French Revolution atheists. Maximilien Robespierre, who presided over much of the Reign of Terror, was a deist who followed the philosopher Jean-Jacques Rousseau. Rousseau, whose views were a sort of simplified Christianity, had no doubts about the existence of God or the immortality of the soul.

In 1794, at Robespierre's urging, the National Convention proclaimed an official religion, the Cult of the Supreme Being, which was based on Rousseau's version of deism. In June of that year, Robespierre led a Festival of the Supreme Being in the Tuileries Garden. Glynn could have got these facts from almost any encyclopedia. He is plainly wrong in ascribing the cruelty of the French Revolution to the atheism of its leaders.

I do not want this to sound flippant, but I know of no example when a group of atheists murdered other people solely because those people held different beliefs. I do not count the Soviet Communists under Stalin, because they were only officially and incidentally atheists. Their crimes were committed for political reasons, and a great many of their followers have now returned to the religions they forsook when it was politically correct to forsake them. Many of the Nazis, by contrast, were devout churchgoers. In addition, to paraphrase Steve Allen, [1993] the well known author, composer, and entertainer, even after 50 years of Communism, there is no evidence that the average person in Communist states has any more favorable a view of robbery, adultery, murder, or rape than do Protestants, Catholics, or Jews in Western countries.

In short, whether or not there is a universal code, all ethics is in the final analysis situational, and neither religion nor a belief in God confers on the believer a higher morality.

* * *

In this context, incidentally, I want to take strong issue with the argument of, for example, the philosopher Brian Leftow, [Morris, 1994] that Christians were not responsible for the

Holocaust or the Inquisition. Rather, says Leftow, these were committed by Gentiles, and German churches were filled with Gentiles. Just attending church services does not make you a Christian, any more than a mouse becomes a cookie by entering a cookie jar.

This is, frankly, one of the worst analogies I have ever encountered, and it is unworthy of a philosopher. The mouse, after all, does not identify itself as a cookie, and the cookies do not recognize the mouse as a cookie. Yet the people who attended German churches were universally recognized as Christians. Those who were responsible for the Inquisition were not "Gentiles... wearing church robes," but rather were the leaders of the Catholic Church in Spain and elsewhere.

Leftow redefines evil Christians as Gentiles. This is an *ad hoc* definition similar to the *ad hoc* hypotheses of astrologers and psychoanalysts (Chapter 2) and is specifically intended to absolve Christianity of responsibility for the evils committed by Christians. If you define evil Christians as Gentiles, then Christians never commit atrocities. Others have probably made the same argument regarding Jews or Muslims. It is no different from arguing that the Americans who committed atrocities in Vietnam were Murcans, not Americans, even though they were American citizens, were products of American upbringing and education, identified themselves as Americans, were culturally Americans, and served in the American army. The only way you can distinguish between a Murcan and an American is to watch their behavior: those who commit atrocities are Murcans. Otherwise, there is no difference.

This is an argument made by a partisan who states explicitly that critics of the New Testament do not have their wits. It is a failed attempt to shelter religion from having to take responsibility for the evils committed in its name.

* * *

Prager and Telushkin go on to argue that without God, which they call a "metaphysical source to life," life must be ultimately purposeless. I suppose, if they mean that the universe would not otherwise have some grand overall purpose, then they are right. But "purpose" does not have to mean external purpose. Our own lives are not necessarily purposeless if there is no external purpose; they have whatever purpose we give them. Even if there is no "metaphysical source to life," I would find it hard to argue that Mohandas K. Gandhi and Martin Luther King, for example, had no purpose in life. (Unfortunately, so did Adolf Hitler and Benito Mussolini.) But more importantly, the argument itself, that God must exist because otherwise life would be purposeless, is no argument. In a novel by Joanne Greenberg, [1964] the principal character's therapist says, "I never promised you a rose garden." Similarly, no one ever promised Prager and Telushkin a universe with some great, eternal plan. All they have is the universe as it exists; if they think it was created with a purpose, then let them at least provide some argument in support of their claim. Instead, they give us no more than wishful thinking, similar to that of the college student.

* * *

Prager and Telushkin conclude their chapter with a number of reasons why some people are atheists. Briefly, they say that atheists believe the way they do because God was presented to them in a childish manner, because they are rebelling against some authority, because they have accepted "ingrained attitudes" that they acquired at home or from their social environment and were not exposed to the arguments of "intelligent believers," or because of their observations of suffering and evil. Though he later apologizes, the philosopher J. P. Moreland [Moreland and Nielsen, 1993] argues similarly that many prominent atheists have had bad relations with their fathers and are rejecting God as a symbol of their rejection of their fathers. These arguments are

95

not so much condescending as irrelevant. Beliefs are not wrong because the believer has a poor reason for believing them or because the believer is in some way reprehensible. That is, the origin of a belief gives no information about its truth or falsity. In the 1930's and 40's, the Nazis rejected Relativity because it was "Jewish physics"; the theory was correct all the same.

More important, though, Prager and Telushkin's arguments can be inverted and applied to believers as well as to nonbelievers: Believers have accepted "ingrained attitudes" that they acquired at home or from their social environment, they are adhering mindlessly to a received dogma, they have not been exposed to the arguments of intelligent atheists, and so on.

More specifically, peer pressure may contribute to belief in God (perhaps less so to a nonbelief in God, since relatively few peers are nonbelievers). I was struck by this passage from the autobiography of a well-known psychologist. [Blackmore, 1986]

> I had been putting myself down, blaming myself, seeing myself as a failure, comparing myself with "successes," doubting everyone and everything and hating myself for it.... I just couldn't cope with the dilemma of psi.... I couldn't believe in psi, but I couldn't *not* believe in psi either. The concept itself seemed nonsensical, but others kept on finding it.

Susan Blackmore, by her own account, wanted desperately to be a parapsychologist. She designed carefully controlled experiments to verify the existence of *psi*, or paranormal powers such as extrasensory perception and telepathy. All her experiments failed. She investigated the experiments of others and invariably found flaws in the experimental design. The complete absence of convincing evidence pointed to the conclusion that psi did not exist. Blackmore, now Professor of Psychology at the University of the West of England, is a fine scientist and well known as a skeptical critic of studies of the paranormal. Yet, because of the beliefs of her friends and colleagues, she held onto her belief in psi long after I thought she should have given it up. Now replace the word "psi" with the word "God" in the excerpt, and change "successes" to "believers." Surely many people believe in God for the same reason that Blackmore continued, against all odds, to believe in psi.

In addition, you can make the case that believers are afraid of death and that their religion or their belief in God is a mere rationalization that allows them to wish it away. (Not all religions posit a god, but all religions have some kind of afterlife, by which I mean reincarnation, resurrection, or immortality of the soul. An afterlife is possibly the only concept shared by all religions. The Western religions link the afterlife to God, and many Americans probably could not believe in the afterlife without God.)

Arguments that purport to explain why people do or do not believe in God—Prager and Telushkin's arguments and their inversions, and the Argument from Peer Pressure—are true in some measure, but they are beside the point. Whether or not there is a God or a metaphysical purpose to existence does not depend on the backgrounds or the motivations of the people who believe or do not believe in it. It is a matter of fact.

Signs

Carl Sagan's novel *Contact* [1986] takes us on a long search but finally leads us to a calculation of the number pi to a great many decimal places. Pi is an irrational number, approximately equal to 3.14159.... It is a string of digits, nearly indistinguishable from a random sequence, that will go on forever without once repeating itself in the sense that 11/27, for example, is the repeating decimal 0.407 407 407.... Nevertheless, in Sagan's novel, the calculation of pi ultimately reveals a string of ones and zeroes that are interpreted as a binary code representing the image of a circle inside a square: a sign from God.

97

If pi were a truly random sequence of infinite length, then it would contain within it every possible finite sequence, much as the proverbial monkeys hacking long enough at a typewriter would eventually generate the works of Shakespeare. To see this, imagine a long sequence of digits arranged randomly. It should be fairly easy to pick a two-digit sequence and find it somewhere within the long sequence. It will be substantially harder to find a three-digit sequence and harder yet to find a four-digit sequence. But an infinitely long sequence is very long indeed, and eventually we will find our three- or four-digit sequence. The length of the random sequence is the key: An infinite sequence is vastly longer than any finite sequence we might want to search for. Therefore, ultimately, we will find any finite sequence whatsoever, as long as the infinite sequence is truly random. Finding a well known text in a truly infinite sequence is not a miracle; it is a certainty.

There would then be no reason to believe that a specific sequence had any significance whatsoever. Indeed, failure to understand this point has led to the discovery of many "signs" where most probably none existed. Pi, however, is not a random sequence, and a discovery such as that in *Contact* would have great significance if it could not otherwise be explained.

Sagan's book was a novel, but it illustrates what I mean by a sign, as opposed to a miracle: a bit of evidence that God plants for us to find at a later date. Such a sign, if it withstood scientific scrutiny, would be very convincing evidence for the existence of God.

Prager and Telushkin [1986] think they may have found such a sign. Perhaps somewhat parochially, they posit the continuing existence of the Jews as a sign from God of his existence. (J. P. Moreland, whom we will meet later (see "The Gospels," Chapter 4), bases a similar argument on the success of the early church against all odds. [Moreland and Nielsen, 1993]) They argue that only the Jews have survived for nearly 4000 years with their culture "intact." I do not know what they mean by intact, but let us agree that the Jewish culture has evolved continuously for nearly that long.

In his book *Return to Sodom and Gomorrah*, the astronomer and archeologist Charles Pellegrino [1994] argues that the Exodus coincided with the eruption of the volcano Thera, which destroyed the Minoan civilization in the Mediterranean. The year of this eruption is known exactly: 1628 B.C.E. If we accept Pellegrino's date, then Jewish civilization is no more than 3600 years old. With the more conventional date of around 1400 B.C.E., it is about 3400 years old. The earliest Chinese dynasty dates from about the same time: 1766 B.C.E. according to one source, and 1523 according to another. [Garraty and Gay, 1972] Thus, the Chinese and Jewish civilizations are around the same age.

In fairness, Prager and Telushkin base their case additionally on the remarkable fact that the Jews have twice lost their homeland and twice got it back, and survived several attempts to destroy them. [See also Keleman, 1990, for more detail.] They surely exaggerate when they claim that the Jews still worship the same God as their earliest ancestors, however; the Jewish concept of God has undergone major revisions since the time of the Exodus (see "The Book of Jonah," Chapter 4).

Some scholars estimate that, in the first century C.E., the number of Jews was roughly 6 million (one-tenth of the population of the Roman Empire), or about equal to the number of Chinese at that time. [Marcus, 1956, p. 114] Today the number of Jews is measured in the tens of millions and the number of Chinese exceeds one billion. By at least some criteria, the Chinese civilization has been by far the more successful.

The ancient Egyptian religion lasted for about 4500 years, nearly 1000 years longer than the Jewish civilization has so far lasted. The first evidence of the Egyptian religion dates to about 4000 B.C.E., and the last temple of the goddess Isis closed in the mid-500's C.E. [Parrinder, 1984, p. 137] It is therefore not unprecedented for a religion to outlive by many centuries the destruction or occupation of its homeland. Thus, without clear evidence of Divine guidance, I do not see how we can reliably describe the survival of the Jews as more than an interesting anomaly, which may have come about either through chance or because of some unknown attribute of the Jewish culture.

* * *

Other attempts to find signs sound like numerology. They are akin to attempts to show that the pyramids were built by space invaders because certain ratios are equal to pi. These are weak arguments in that the authors search at random for "meaningful" ratios and ignore ratios that are not meaningful; that is, they use evidence selectively. You can find one such argument in *Genesis and the Big Bang* by Gerald Schroeder [1990]. Schroeder is a physicist who attempts to relate the creation myth of the Book of Genesis to modern physics. For example, Schroeder describes relativistic time dilation in layman's terms and then states flatly that the six days of creation are six days when viewed from the proper frame of reference but are billions of years when viewed from earth. He provides no argument to support this claim but rather seems to think that stating it is enough. Indeed, his preconception is made obvious from the outset, when he makes clear that his intention is to reconcile the creation myth with modern science, not to examine the truth of either. (First published in [Young, 1998a].)

In an appendix, Schroeder presents a typical numerological argument: Take the Hebrew Bible. Go to the first instance of the letter *tav*. Count 50 letters (49 spaces) and write the fiftieth letter. Count 50 letters again and yet again, and you get the Hebrew word *TVRH*, or *Torah* (the five books of Moses). Do the same thing with the Book of Exodus, the second book of the Hebrew Bible, and you get the same result. Now take Deuteronomy and Numbers, the fourth and fifth books, and you will get *TVRH* spelled backward.

In Leviticus, the central book, you use a different rule. Go to the first *yod* and count eight letters (seven spaces) and write the eighth letter. Count eight more letters and yet eight more. You get the letters *YHVH*, the name of God in Hebrew. Thus, the four words *TVRH* in the first and last two books point toward the name of God in the central book.

Why count 49 spaces? Because we count 49 days from Passover to Shavuot (Pentecost) and celebrate Shavuot on the fiftieth day. What is the connection? I haven't the foggiest idea, and Schroeder hints at none.

Furthermore, the procedure is not as cut and dried as Schroeder claims. In Genesis and Exodus, the rule works. In Numbers, however, you do not begin from the first *heh* but rather from the third (remember that *TVRH* is spelled backward). In Deuteronomy, you have to begin counting in the fifth verse, and you count only 48 spaces, not 49. It looks suspiciously as if you get to make up the rules as you go along.

By the same token, consider Psalm 46 in the King James Version. Ignore the dedication and begin with the words, "God is our refuge and strength." Count 46 words, and you will find the word *shake*. Go to the end and ignore the concluding word, "Selah." Count backward 46 words, and you will find *spear*. Put them together, and you get "Shakespear," which is a legitimate alternate spelling. Are we to conclude then that the works of Shakespeare were dictated by God and add them to the Bible? No. Better to conclude that Schroeder is dredging for correlations (or mining his data) and incorrectly ascribing significance to correlations that occur by chance alone.

I am reminded of a statement attributed to the physicist Richard Feynman: "Wow! I just saw a car with the license number MQL7243. Do you know what the odds against that are?!" What Feynman is saying is that you can always find a coincidence that retrospectively looks highly improbable. Such a coincidence is proof of nothing unless you can predict it ahead of time.

* * *

A far more interesting attempt to prove Divine authorship embodies a concept known as *equidistant letter sequences*. [Ben-David, 1995; Satinover, 1995] This involves writing the Book of Genesis (in Hebrew) without spaces and in a font that does not use proportional spacing. It is as if we wrote the first eight verses of the King James Version as shown in Figure 2. Then we search for words embedded in the text, much in the manner of a word hunt puzzle. The words, however, need not be restricted to vertical or sloped at 45 degrees, but can occur at almost any angle. For example, you might start with one letter, then skip 12 letters to get the second, skip 12 more to get the third, and so on. The length of the line is therefore irrelevant.

```
INTHEBEGINNINGGODCREATEDTHEHEAVENANDTHEEARTHANDTHEEART
HWASWITHOUTFORMANDVOIDANDDARKNESSWASUPONTHEFACEOFTHEDE
EPANDTHESPIRITOFGODMOVEDUPONTHEFACEOFTHEWATERSANDGODSA
IDLETTHEREBELIGHTANDTHEREWASLIGHTANDGODSAWTHELIGHTTHAT
ITWASGOODANDGODDIVIDEDTHELIGHTFROMTHEDARKNESSANDGODCAL
LEDTHELIGHTDAYANDTHEDARKNESSHECALLEDNIGHTANDTHEEVENING
ANDTHEMORNINGWERETHEFIRSTDAYANDGODSAIDLETTHEREBEAFIRMA
MENTINTHEMIDSTOFTHEWATERSANDLETITDIVIDETHEWATERSFROMTH
EWATERSANDGODMADETHEFIRMAMENTANDDIVIDEDTHEWATERSWHICHW
EREUNDERTHEFIRMAMENTFROMTHEWATERSWHICHWEREABOVETHEFIRM
AMENTANDITWASSOANDGODCALLEDTHEFIRMAMENTHEAVENANDTHEEVE
NINGANDTHEMORNINGWERETHESECONDDAY
```

Figure 2. Dredging for data in the first eight verses of the King James Version of the Bible. Circled words are interpreted, whereas words in **boldface** are ignored.

Using my eye (and a ruler), I found about 25 embedded words in a matter of minutes. For example, near the left edge, you will find a vertical *ten*, the number of commandments. This word happens to run vertically, but you could describe it as *t*, 53 spaces, *e*, 53 spaces, *n*; hence the term equidistant letter sequences. Note further that the 53 letter *skip* does not depend on the length of the lines but would still be 53 letters if I shortened or lengthened the lines. If, for example, I made the lines 52 or 54 characters long, then the word *ten* would run downward at 45 degrees. At any rate, near the upper left, running upward, you will find *pun*, an indication that the Hebrew Bible is full of plays on words; *lion*, as in Lion of Judah; and *Lot*, Abraham's nephew. A little to the left of *ten* is *rent*, as in *rent his garments*. At right is *hasid*, the Hebrew word for pious or ultra-orthodox, which proves that the ultra-orthodox are right. On the other hand, near the center, we find *Leda*, as in Leda and the swan, which shows that all of Greek mythology is also correct. Just above and to the right of center, you will even see *HeNe*, which shows that the Bible anticipates the invention of the helium-neon laser.

So far, I am doing numerology in that I selectively chose the words I liked and ignored words like feet, lame, halt, near, hot, and art. That is, like Gerald Schroeder, I am dredging for correlations. That is almost certainly what Michael Drosnin [1997a] does in his best seller *The Bible Codes*: he dredges for correlations. [Horovitz, 1997; Hendel, 1997; Odenheimer, 1997] Drosnin, a journalist, looks for pairs of words, not just isolated words, at different skips. He claims that he can predict the future by evaluating such pairs; most spectacularly, he claims to have predicted the assassination of Yitzhak Rabin by correlating Rabin's name with the words "assassin that will assassinate."

David Thomas, [1997] a physicist, shows how easy it is to dredge for such correlations. He uses the King James Version and easily finds Roswell and UFO (unidentified flying object) embedded in Genesis, for example. (Roswell, New Mexico, is the site where an Air Force surveillance balloon crash landed and was years later misidentified as a craft from outer space, or a UFO.) Has the Bible (in translation, no less) predicted the appearance of UFO's?

Thomas searches Genesis for the names of modern personages such as Dole, Leno, Newt, Clinton, Kennedy, and Einstein. He chose five each of four-, five-, six-, seven-, and

103

eight- or nine-letter names and searched for them in the King James Version of Genesis. He found thousands of instances of the four-letter names he chose, dozens of some of the five-letter names, a few of the six-letter names, and none whatsoever of the seven-, eight-, or nine-letter names.

Thomas searched forward only and limited his skips to numbers less than or equal to 1000. Had he used longer skips and searched backward, he would have found longer names as well. Nevertheless, his failure to find longer names underlines an important point: Short words are easier to find than long words. Because Hebrew may be written with few or no vowels, Hebrew words are shorter than English words, so it is easier to find embedded words in Hebrew than in English. Additionally, we have to ask why God would have failed to encode Einstein or Churchill, at least with relatively short skips, while carefully encoding Oprah 49 times and (Ralph?) Reed 7340 times. Oprah, by the way, is Harpo spelled backward; which rendering did God intend?

Drosnin has challenged anyone to find paired messages in any book, such as *War and Peace*. Thomas took up the challenge and searched the first 170,000 characters of *War and Peace*; that is approximately the same number of characters as in the King James Version of Genesis. Thomas easily found paired messages in that and every other English text he worked with. For example, he wrote a program to find "Hitler" and "Nazi," and found these words a half-dozen times in his selection from *War and Peace*. It is startlingly easy to find hidden "messages" in any text; the question is whether they mean anything.

Rabin's name appears in Genesis (in Hebrew) with a skip of 4772, according to Shlomo Sternberg, [1997] an Orthodox rabbi and professor of mathematics at Harvard. If you arrange the letters correctly, that is, with the right line length, you indeed see the words Yitzhak Rabin running vertically and crossing Deuteronomy 4:42,

104

[41] Then Moses set aside three cities on the east side
of the Jordan, [42] to which a manslayer could escape,
one who unwittingly slew a fellow man without
having been hostile to him in the past.... [literally,
without knowledge and he did not hate him]

What the Jewish Publication Society renders "manslayer
who slew," Drosnin translates as "assassin that will assassinate,"
which is a reasonable translation if the words are taken wholly
out of context. Sternberg's translation, "a slayer who happens to
have killed," is more in keeping with the sentiment that the
killing was unintentional and without malice.

But let us look at the verse as a whole, rather than pick out
three isolated words. It refers to a man who unknowingly kills a
fellow man *whom he has nothing against.* Even if you believe
Yigal Amir's claim that he intended only to wound Rabin, it
would take a wild leap to argue that Amir held nothing against
Rabin and that the killing was wholly unintentional. You cannot
shoot someone and claim that you had nothing against him.
Thus, even if the correlation were significant, it would be hard to
see how the verses in question refer to Rabin's assassination.

In addition, a simple correlation between "Yitzhak Rabin"
and "assassin who will assassinate" tells us nothing quantifiable.
Will an assassin assassinate Rabin? Or is Rabin an assassin
himself? Ancient Hebrew did not have tenses in the way we
think of tenses in modern English. [Chomsky, 1957] Instead,
ancient Hebrew used completed action and continuing action.
Continuing action evolved into the future tense and often
corresponded with the future, but not always. Consider, for
example, the story of Hannah, the favored but childless wife (1
Samuel). Elkanah's wife, Penninah, bore children, but Hannah
had none, even though Elkanah favored Hannah.

[6] And her rival [Penninah] also provoked her...
because the Lord had shut up her womb. [7] And as
he [the Lord] did so year by year, when she went up

> to the house of the Lord, so she provoked her;
> therefore she [Hannah] wept, and did not eat. [8] Then
> said Elkanah her husband to her, Hannah, why
> weepest thou? and why eatest thou not?

The Hebrew words for *weepest* and *eatest* are *tivki* and *tochli*. Out of context, they would be translated as "you will weep" and "you will eat." But here these translations make no sense, inasmuch as Hannah is already weeping. Instead, the meaning is, "Why do you *keep on* crying? Why do you *keep on* not eating?"

The words "assassin who will assassinate" similarly imply continuing action and could as well be translated as "assassin who *is continuing* to assassinate" or "assassin who *keeps on* assassinating." I have no doubt that a Muslim journalist dredging for Bible codes would argue that Rabin, as an Israeli general, kept on assassinating, that is, that Rabin was the assassin, not the victim, to whom the code refers.

Drosnin was severely criticized in *The Jerusalem Report*, which could find no academic support for Drosnin whatsoever. Even Drosnin's mentor, Eliyahu Rips, whom we will meet directly below, derides Drosnin's failure to show that his correlations are not the result of chance.

It is interesting that Jewish literalists do not accept Drosnin's claim that he can use the codes to make predictions. They argue that the codes are planted only to refer to events past; since there is no real syntax, you cannot interpret the codes as anything other than signs that God has planted. Their argument is based on the Biblical injunction (Deuteronomy 18:10-12)

> [10] There shall not be found among you any one that...
> useth divination, or an observer of times, or an
> enchanter, or a witch, [11] Or a charmer, or a consulter
> with familiar spirits, or a wizard, or a necromancer.
> [12] For all that do these things are an abomination
> unto the Lord.

Drosnin [1997b] himself argues that no one has proven that the codes are not real. He has found "Yitzhak Rabin" encoded in the Hebrew Bible only once and seems to think the fact is significant.   Further, he found the correlation before the assassination and warned Rabin a year ahead of the event. He was not taken seriously.

* * *

If I wanted to show that the embedded words have real meaning, I would have to show that they appear in the text with greater frequency than would be expected by chance. That is precisely what Eliyahu Rips of the Hebrew University of Jerusalem and his colleagues Doron Witztum and Yoav Rosenberg have attempted to do. [Witztum, Rips, and Rosenberg, 1994] They published their results in the peer-reviewed journal *Statistical Science*.

Witztum and his colleagues first looked for a smallish number of embedded words. Then they searched for words that had related meanings to see whether they appeared relatively close to one another in the text. For example, if they found *hammer*, then they searched nearby for *anvil*. If they found *Chanukah*, the festival that marks the capture of the Temple from the Syrians, then they looked for *Hasmonean*, the name of the dynasty established after the rededication of the Temple.

Well aware that such correlations could come about as the result of chance, Witztum and his colleagues performed another experiment, in which the pairs they searched for were the names of historical figures and their birth dates. The historical figures were eminent rabbis. Their names and the dates were taken from a standard Hebrew reference work and chosen according to the length of the citation; only the figures with the longest citations qualified.   All the historical figures lived long after the completion of the Book of Genesis, so they could not have been known to its authors. Whenever Witztum and colleagues found the name of one of the historical figures, they searched in the vicinity of that name for his birth date. They found birth dates in the vicinity of the names with greater frequency than was predicted by chance. This result seemed so bizarre to the referees of their paper that they insisted that the authors perform other tests, such as rearranging the letters randomly and looking again for correlations. The authors complied but found no correlations in the randomized texts. The probability that their results could have happened by chance is about 2 in 100,000. A statistician I consulted could find no obvious errors in their paper and was as baffled as the referees.

Jeffrey B. Satinover, [1995] a psychiatrist in private practice, describes the work of Witztum and colleagues in *Bible Review*. Satinover confuses the search for the pairs of related words with the search for the rabbis' names and dates, and he overstates the results. The authors did not find every word pair they looked for; they found only a statistical regularity that exceeded what they had expected from chance alone. Satinover also quotes odds of approximately 1 in 10^{17}, even though this number does not appear in the *Statistical Science* paper. He further remarks that believers automatically accept the results, whereas skeptics automatically reject them. Without quoting anyone by name, he says that scientifically trained skeptics have been asked, "What standard of proof would you accept... that the phenomenon might be genuine?" They reply, "There is no standard. I will not believe it regardless."

I do not consider that a fair response. I think it is, however, fair to say that this result is an interesting anomaly that bears further scrutiny. According to a box attached to Ben-David's article, Moshe Zeldman, a rabbi with training in mathematics, argues that God might have "wanted [the equidistant letter sequences] discovered in an age of scientific skepticism about divine authorship." That proposition is just too bizarre and too anthropomorphic for many who do not already believe in divine authorship. This is not the first skeptical generation. Indeed, churchgoing in the U.S. and, possibly, religious observance in the world as a whole are on the rise. Why did God not put some in some clues for the generation of the 1950's, when religious attendance was at a low point? Why does God have to beat around the bush? Why does he not simply offer obvious and incontrovertible proof to each generation?

The embedded codes are cryptic at best—a word or two here, a word or two there. In *Contact*, God encodes enough data to form a circle inside a square. Though Sagan does not specify, presumably it would take at least several hundred digits to draw a convincing circle on a computer monitor. If God, who has undoubtedly read *Contact*, wanted to convince us unequivocally, then why did he not encode a substantive message or an unambiguous prediction, instead of playing parlor games and encoding ambiguous messages of no more than a few letters each? At the risk of anthropomorphizing, may I suggest that, if God wanted to be really convincing, he would have gone straight to the first *aleph* in Genesis (the third letter) and encoded a significant statement such as "I am the Lord your God who brought you out of the land of Egypt" (Exodus 20:2). That is 32 letters in Hebrew. To be more convincing, he would have made the skip significant as well: 49 spaces, for example. It is considerations like these, not the rigidity implied by Satinover, that made skeptics initially skeptical.

My friend Stanley, a physicist, thinks that Witztum's result can be explained away by noting that a book is not a random sequence of letters, but rather is an ordered sequence. Stanley explains that the first letter in the King James Version is *i*. The probability that the next letter is *n* is very high, since *in* is a very common word or prefix in English. Given that the first two letters are *i* and *n*, the probability that the next letter is *t* is also high, since *in the* is a common phrase in English. Thus, if the first letter is *i*, the odds are fairly good that the fifth letter will be *e*. The next (sixth) letter, however, could be almost anything: in the beginning, in the last analysis, in the morning, in the zoo, and so on. Thus, if the skip is more than five or ten letters, the probability of finding a given letter becomes virtually indistinguishable from that of a random sequence. Stanley's hypothesis is falsified.

My friend Jared dismissed Gerald Schroeder's numerology by calling it "very clever." Jared, a former rabbinical student, has attended countless sessions where the rabbi indulged in clever numerology similar to Schroeder's, and he is unsympathetic with Witztum's work as well. He suspects that Witztum and his colleagues have simply found something that looks improbable and calculated its probability, much as Feynman was struck by the license plate number. Specifically, Jared says, consider the probability that six cars pass my window and their colors form the sequence blue, maroon, tan, black, green, blue. What are the odds against that sequence occurring? Probably fairly small—yet that is precisely what has just happened. If Jared had predicted this sequence before seeing the cars, then that would have been impressive, but *retrodicting* it and then calculating the probability is meaningless. It is very easy to assign vanishingly small probabilities to events that have actually occurred, but you have proven nothing by doing so. Jared suspects Witztum and his colleagues of retrodicting in this way.

In addition, Brendan McKay, Professor of Computer Science at the Australian National University, has pointed out to me that, if Jared had really been able to predict the sequence of cars, I would have been impressed with Jared, whereas Witztum would have been impressed with the cars. That is, even if there are regularities in the Book of Genesis, finding them may say more about Witztum's insight into the structure of language than it says about Genesis itself.

I initially found it hard to accept Jared's argument, however, since the names of the historical figures were supposedly chosen by a fairly complex algorithm and not picked precisely because they worked. It now seems, however, that Jared was right, and I bow to his greater insight. According to an exchange of mail on the Internet and an article in a Hebrew-language science magazine, [Bar-Hillel, Bar-Natan, and McKay, 1997] Witztum and colleagues have dredged for correlations by adjusting the input parameters in such a way as to create the result they wanted.

How? Maya Bar-Hillel of the Center for the Study of Rationality at the Hebrew University of Jerusalem and her colleagues, mathematicians Dror Bar-Natan of the Hebrew University and Brendan McKay, point out that people are often called by more than one descriptor. For example, Robert F. Kennedy may have been called Bobby Kennedy, Robert Kennedy, Senator Kennedy, Sen. Kennedy, RFK, or Robert Francis Kennedy. If you wanted to dredge for correlations, you would search for each of these but include only the one that worked best when you estimated the odds.

In addition, Hebrew spelling was not standardized until very recently, so we know many of the eminent rabbis by several variant spellings, much as we know Shakespeare as Shakespear and Shakspere, as well as by other variants. The name Jonathan, for example, appears in the Bible as *YVNTN* and as *YHVNTN*; you could search for both and ignore the spelling that did not work. Similarly, beloved rabbis usually have nicknames. Often these nicknames are acronyms. Thus, *Rav Sh*lomo ben *Y*itzchak (Rabbi Solomon son of Isaac) is most commonly known as *Rashi*. If you wanted to dredge for correlations, you could search for Shlomo, Rav Shlomo, Rav Shlomo ben Yitzchak, Shlomo ben Yitzchak, or Rashi. Again, you would ignore the variants that did not work.

Witztum and his colleagues used three ways to express the dates: In Hebrew, letters substitute for numerals, so you can find dates embedded in a text as easily as you can find words or names. If a rabbi was born on the fourth of the month Tevet, you may express that as *4 Tevet*, *on 4 Tevet*, or *4 in Tevet*, where 4 is indicated by the fourth letter *dalet*. Compare these with 4 July, on the fourth of July, and the fourth of July. Witztum and his colleagues used all three descriptors, thereby increasing their odds of finding a match. They did not use the fourth possibility, *on 4 in Tevet*.

111

Bar-Hillel and her colleagues performed the same calculations as Witztum, except that they systematically varied the spellings of the rabbis' names and the form of the dates. They got the best correlations when they used the names and dates in the forms chosen—supposedly *a priori*—by Witztum and his colleagues. That is, as Bar-Hillel and her colleagues expressed it, "... wonder of wonders, it turns out that almost always, if not always, the allegedly blind choices paid off: just about anything that could have been done differently than it was actually done would have been detrimental to the [result]."

Perhaps we cannot dismiss the work of Witztum and his colleagues quite as cavalierly as Jared does, but it certainly seems probable that, consciously or otherwise, they have finagled in order to get the "right" result. In addition, as I will show below, there is a great deal of evidence that the Bible was assembled, or edited, rather than written as a single work. To believe Witztum's interpretation uncritically, we would have to believe that God guided the hands of countless scribes over many generations in order to get precisely the right text to prove his point to us, over 1000 years later, and ignored intervening generations entirely.

Conclusion

None of the popular arguments in this chapter survives scientific scrutiny. Coincidences are claimed as miracles. God is excused for every misfortune. *Ad hoc* hypotheses abound: God allows evil in order to give us free will. True believers are not evil, even though many evil people profess to be true believers. There must be a God because otherwise our existence would be purposeless. Equidistant letter sequences are claimed as evidence that God wrote the Bible, but they are based on statistics that cannot stand scrutiny and may have been finagled. All these arguments rely on *ad hoc* hypotheses—fixes—by adherents whose minds are made up and who are uninfluenced by the need for evidence. Not one meets an acceptable standard of proof.

In the next chapter, we will continue our examination, now asking whether the Bible has a legitimate claim to historical accuracy.

We do not reject the eternity of the universe because
certain passages in Scripture confirm the creation,
for... it is not difficult to find for them a suitable
interpretation.

RAMBAM
(MOSES MAIMONIDES)

Believe those who are seeking the truth; doubt those
who find it.

ANDRÉ GIDE

[T]he means of studying nature is what men call
science. It is therefore absurd for any religious
believer to perceive himself as defending God by
attacking science.

STEVE ALLEN

Those who fear the facts will forever try to discredit
the fact-finders.

DANIEL DENNETT

Chapter 4
Questioning authority

Menachem Mendel Schneerson, the Lubavitcher rebbe, was
the spiritual leader of a few tens of thousands of Chassidim, or
ultra-Orthodox Jews. (A rebbe is a Chassidic rabbi; the word is
a dialectical pronunciation of rabbi.) It would be an
understatement to call Schneerson the Pope of the Lubavitcher
Chassidim. He was more accessible to his followers and gave
advice on secular matters as well as religious. Born in 1902,
Schneerson studied at the University of Berlin and received a
degree in engineering from the Sorbonne in the early part of the
twentieth century. He and his followers emigrated to the United

States in the mid-thirties and settled in Brooklyn. Schneerson died in 1994. His successor has not been named.

In the mid-sixties, according to the journalist Lis Harris [1985], Schneerson wrote a letter to a follower who was concerned about differences between what modern science taught and what the Torah (no doubt as interpreted by the Lubavitcher rebbe) taught. The Torah is, literally, what Jews call the first five books of Moses, but many Jews use "Torah" to mean the entire Hebrew Bible as well as the later writings called the Talmud. Schneerson's letter does not make clear in which sense he means "Torah," but I think we can assume from the context that he is writing mostly about the Biblical account of the creation.

I think of Schneerson's letter as a sort of Papal encyclical, and Harris calls it the Lubavitchers' "polestar" on scientific matters. Schneerson tells his follower that reconciling science and Torah is a hopeless task, because, as quoted by Harris, "science... deals with theories and hypotheses, while the Torah deals with absolute truths." Schneerson argues that there is no *evidence* [italics in Harris's text] in favor of evolution, since no one has ever seen a single species "transmute" into another species; hence, the theory must not be empirical. Like other creationists, Schneerson argues further that God could have put the fossils into the ground for some reason that we cannot know, or that the conditions on earth could have been so different in the past that in fact the fossils were formed within the last 5000 years. He evidently dismisses the idea, which is perfectly reasonable from a religious perspective, that God allowed the fossils to form so that we would one day be able to investigate our distant past. (Additionally, I wonder how Schneerson would respond to the analogous argument that the world was created in 100 C.E. and that therefore the Torah was created already written and describes events that never happened. If God could have created the fossils in that way, why not the Torah? [Napier, 1999])

What Schneerson is saying is that scientific conclusions are inferences—perhaps I should say *mere* inferences—whereas the Torah is absolutely true.

It is almost impossible to respond to this argument. Schneerson's odd use of "empirical" implies that the only valid evidence is direct evidence. His claim that the conclusions of science are inferences is correct; all knowledge is in some sense an inference. But the statement that the Torah is absolutely true is *not even* an inference. It is a flat statement without any argument whatsoever in its support. To return to an earlier example, it is a little like telling me that the moon is made of green cheese and never mind that scientists say it is composed of rock. Whether or not the Torah (or any other religious document) is or is not absolutely true, as I have noted before, is a matter of fact, and we ought therefore to be able to examine the claim intellectually. Schneerson seems unwilling or unable to do so.

In a 1998 speech in Boulder, David Saperstein, a Reform rabbi, argued that literalists such as Schneerson believe in a severely limited God: Schneerson's God is incapable of writing allegory. Therefore literalists must believe in the creation account literally. It is perhaps a slight overstatement; Jewish literalists, for example, consider the Song of Songs an allegory for the love between God and Israel, whereas to me it is no more and no less than love poetry. The Song of Songs is different from the Torah, however, because it was supposedly written by Solomon. Apparently Solomon could write allegory, but not God.

* * *

In any case, Schneerson's claim that no one has ever seen species transmute is not wholly accurate. Carl Sagan, the well known astrophysicist, had formal training in biology and studied evolution and the origin of life as well. In *The Dragons of Eden*, [1977] Sagan cites the widely known example of certain English

117

moths that cling to the bark of white birch trees. The moths used to be white; their color matched the bark of the birch trees and gave them camouflage from predators. Because of accidental mutations, some of these moths produced enough of the pigment melanin to color them black; black moths were usually eaten by predators and did not survive long enough to reproduce. That is, natural selection worked against the survival of the black moths, and they remained a small fraction of the population.

After the Industrial Revolution, however, soot covered the birch trees and made them black. The white moths became starkly visible against the now black trunks of the trees and were eaten by predators. The black moths, those with the mutation, were camouflaged. In consequence, they survived and reproduced in greater numbers than the white moths. Virtually all the moths are black now, though there are occasional mutations to the white color.

The black moths are not a different species from the white moths but rather a different variety or subspecies. With time, they could become a different species from another geographically isolated population of white moths that have not been influenced by the Industrial Revolution. Whether or not the black and white moths are different species, the change of the moths from white to black as the environment changed is a clear example of natural selection in action.

Another example is Peter and Rosemary Grant's studies of the finches on the Galápagos Islands; the Grants, both of Princeton University, watched over a 20 year period as the beaks of the finches changed as the availability of seeds changed in response to changes in climate. [Weiner, 1995] Specifically, during a drought, the seeds on which the finches fed grew hard and difficult to crack. Only finches with strong, blunt beaks survived, and they passed their strong beaks on to their offspring. The character of the beak can change within a single generation. The Grants are further observing rare interspecies breeding, which can result in new species under certain conditions.

118

Finally, the apple is an import from Europe. The apple maggot is the larva of a certain kind of fruit fly. The apple maggot today feeds on apples, but it originally fed on the fruit of the hawthorn tree. [Gibbons, 1996] Only a few years after the publication of *Origin of Species*, the entomologist Benjamin Walsh suggested that the maggots that fed on apples had become a different species from those that fed on hawthorn fruits. The usual way to distinguish between two species is to determine whether they normally interbreed; if they do not, then they are, by definition, different species. Walsh could not distinguish between the fruit flies that laid eggs on hawthorn fruits and those that laid eggs on apples, so the issue was not resolved.

In the 1980's, however, Guy Bush of Michigan State University noticed that certain male fruit flies performed a courtship dance on the fruits of the apple tree, whereas other males danced on the hawthorn fruits. He speculated that some fruit flies would mate only on apples and others, only on hawthorn fruits. If that were so, then the two cohorts of flies were different species. To test his hypothesis, Bush and his colleague Ron Prokopy watched flies for hours and concluded that apple flies mated almost exclusively with other apple flies, whereas hawthorn flies mated with hawthorn flies. Bush's former graduate student Jeff Feder, now at the University of Notre Dame, dyed fruit flies in order to identify them more clearly and found that only 6 percent of them cross-bred. It is probable that the two cohorts of fruit flies no longer interbreed because the adult flies emerge at different times that coincide with the times at which the trees (apples and hawthorns) set fruit. Bush, Feder, and their coworkers have studied other fruit flies that are closely related to the apple flies and have the same range but feed instead on blueberries. They hypothesize that the two species, which are now clearly distinct, may have diverged after adapting to specific plants. Thus, it is not quite true that we have not seen the divergence of species; we are clearly seeing the beginning of the process and may have seen, in the blueberry flies, an actual example.

119

Not only the evidence I sketched above but also the well known examples of resistance to herbicides, insecticides, and antibiotics provide overwhelming evidence in favor of intraspecies variation over time. Thus it is no surprise that those who refuse to believe in evolution now admit, like the Lubavitcher rebbe, to seeing intraspecies variation, even though they deny that it can lead to speciation. This is nevertheless a welcome development, since it is a retreat on their part as the ground on which they base their belief is slowly eroded away.

The Bible as a science text

Gerald Schroeder, whom we met in connection with the Bible codes, descends from a different stream of Jewish literalism than the Lubavitcher rebbe. In *The Science of God* [1997], he states at the outset that he will examine modern scientific texts and ancient religious texts and try to bring them into agreement. If that sounds sensible to you, try turning it around: What if he were going to limit his study to modern theological texts and ancient scientific texts and try to bring them into agreement? You would very possibly recommend that he extend his study to include modern scientific texts. But Schroeder evinces no interest in religious studies later than those of Nachmanides (1194-1270). (First published in [Young, 1998b].)

If you go into an investigation with what I will call a thesis, you will very probably prove that thesis to your satisfaction, whether or not the evidence would be convincing to a more or less neutral observer. Schroeder believes that the account of the creation in the Bible is literally true, and he is determined to prove it. In support of this thesis, he latches very firmly onto Big Bang theory but rejects evolutionary biology. To my mind, Big Bang theory is the more suspect. Although it is probably right in outline, it is in the last analysis an extrapolation to densities and temperatures at which our physics has simply never been tested. Extrapolating General Relativity too far back in time and therefore density may be like extending a gas law to temperatures or pressures at which the substance has liquefied. The law breaks down and becomes completely inapplicable. Evolutionary biology, by contrast, is established fact, even if there is yet some controversy about the details. Ultimately, as we will see, not even Schroeder can escape the fact of evolution.

Schroeder believes that the first chapter of Genesis accurately describes the creation as viewed from the proper relativistic frame of reference. He calls time as seen from this frame of reference *cosmic time.* Cosmic time is not entirely arbitrary but is based on a philosophical argument by the theoretical physicist Jean-Marc Lévy-Leblond. [1990] Lévy-Leblond argues that the finite age of the universe puzzles not only laypersons but also scientists and philosophers. To eliminate the conceptual ambiguity that arises when we claim that time had a beginning, Lévy-Leblond defines what he calls *linear time.* Linear time is set up precisely so that the age of the universe is infinite. If you measure time in linear time, the question, "What came before the beginning of time?" is meaningless.

Schroeder renames linear time cosmic time. According to Schroeder, Chapter 1 of Genesis was written from the point of view of a being living in cosmic time. Only at the end of Chapter 1 is the clock turned over to earthly beings. Specifically, Schroeder argues that the first day of Genesis corresponds to the first 8 billion years of earth time, the second day to 4 billion years, and so on.

The correlation between cosmological and paleontological fact and the six days of creation is fairly good, but there are problems. In the Book of Genesis, there is water above the sky; dry land and plants are created before the sun; and fowl before reptiles. Schroeder notes that flowering plants would not have appeared on the third day but on the fifth, but solves his dilemma with an *ad hoc* argument taken from Nachmanides: the plants *developed* during the next days. What did they develop from? Single-celled plants. Sounds a good bit like evolution.

121

Similarly, there is no geological evidence for a Noachian flood, so Schroeder applies a textual argument to suggest that the flood was only local. In brief, he notes a change in terminology from *eretz* to *adamah*. These words, in context, are synonyms, but to Schroeder they signify that God changed his mind about a worldwide flood. Finally, at the beginning of the sixth day, there should have been a major mass extinction, but it is not mentioned in Genesis, and Schroeder all but ignores it.

Like the Lubavitcher rebbe, Schroeder admits to microevolution but not macroevolution (speciation). To "prove" that speciation is not possible, Schroeder carries out a back-of-the-envelope calculation to suggest that the required number of mutations could not take place in a short enough time. The calculation is flawed by the assumption of independent probabilities, but it has other problems as well.

Schroeder is correct in that some biologists argue that evolution takes place as the result of fortuitous mutations. Nevertheless, others argue that at least some speciation is the result of what Schroeder calls microevolution operating on isolated or stressed populations over generations. Nilsson and Pelger's computer simulations, which are far more sophisticated than any back-of-the-envelope calculation, suggest that an eye can form from a light-sensitive spot within a decidedly finite time. Schroeder is nobody's fool, and he surely understands evolution, so I find it hard to believe that he is wholly ingenuous describing evolution as simplistically he does.

Schroeder seems positively triumphant when he discovers that eyes may not have evolved independently several times. In fact, the development of eyes in animals as distinct as vertebrates, cephalopods, and insects may be governed by the same gene, a gene known as *Pax-6*. [Zuker, 1994] If that is so, then eyes may not have evolved independently in different phyla, as has been commonly thought. Michael Grant, Professor of Biology at the University of Colorado, tells me, however, that the existence of a common early gene (*Pax-6* in this case) does not necessarily demonstrate common ancestry and that you can still make a sensible argument that eyes developed differently in different phyla, even though they all use the *Pax-6* gene. (I immediately thought of different cultures using mud to make bricks, even though they had no contact with each other.) Indeed, the differences in the structures and the developmental pathways among eyes in different phyla suggest that the eyeballs themselves do not share a common origin. As I noted earlier (see "Taste," Chapter 2), the layer of blood vessels lies above the nerve layer in the mammalian eye, but not in all eyes. The relation of the *Pax-6* gene to the development of the eye is therefore by no means settled, and it may take years before a consensus is reached.

Schroeder uses the discovery of the *Pax-6* gene to suggest that evolution has been directed in the sense that these gene clusters were prepared for precisely the purposes to which nature later put them: a typical God-of-the-gaps argument (see "The Argument from First Cause," Chapter 6). Unfortunately, he does not develop the argument further, but rather is content with knocking down straw men with back-of-the-envelope calculations. Even Schroeder cannot escape the fossil record, however, and he and indulges in a form of evolution, even while crying the opposite. Indeed, Schroeder is performing a service in that he may be helping to drag other Biblical literalists into the twentieth century; it would be churlish of me to complain that it is only the first half of the century.

Schroeder fixes on the supposed absence of transitional forms in the fossil record and suggests that such gaps disprove evolution. One transitional form that excites him, however, is Archeopteryx, a birdlike creature with teeth. He associates Archeopteryx with *tinshemet*, which is categorized both as a bird (Leviticus 11:18) and as a reptile (11:30). The Jewish Publication Society renders the two instances as *horned owl* and *chameleon* but notes that many of the words in this section have been lost and their meanings are uncertain. Eliding quickly from speculation to fact, Schroeder informs us that an understanding of Archeopteryx/*tinshemet* may provide an insight equivalent to

123

the Rosetta stone. He is, however, mistaken in thinking that there are no other intermediate forms. In fact, there are many, as long as you do not demand an intermediate between distantly related species, like horse and human. Indeed, the progression from Eohippus to modern horse, which is depicted in many textbooks, shows a series of intermediate forms. If you examine the DNA of different species, you find practically a continuum of intermediate forms.

The Hebrew word for soul is *neshamah*. (It may have the same root as *tinshemet*; what are we to make of that?) Schroeder believes that Adam and Eve were the first humans in the sense that they were the first hominids to be endowed with a *neshamah*. He does not deny the existence of prehuman and therefore pre-Adamic hominids whose fossils are indistinguishable from those of human hominids. Only after hominids were given souls, however, did they develop writing and civilization. Part of the evidence is from the language of the Bible. For example, Genesis 2:7, "And the Lord God formed man of the dust of the ground," spells "formed" *VYYZR* (*va-yiytzer*; Schroeder incorrectly transcribes it as *ya-tsar*), whereas Genesis 2:19, "And out of the ground the Lord God formed every beast of the field," spells it *VYZR* (*va-yitzer*). The Hebrew letter *Y* (*yud*) is the first letter in *YHVH*, the name of God, and is often used as an abbreviation for God. It is put into the verb in 2:7 to show that humans have an extra spiritual input. If you had asked me, I might have noted that the Bible was written over a long time and then copied over and over. Post-Biblical Hebrew was developing a system of using *Y* and *V* as vowels, and possibly one scribe or another wrote the first instance of *VYYZR* with the second *Y* to signify the vowel. Contrary to what you commonly hear, the Hebrew Bible was not stabilized until the invention of the printing press (see "The Hebrew Bible," this chapter). But Schroeder has no apparent interest in modern Torah scholarship. Instead, he gives deep significance to what are probably no more than scribal errors or inconsistencies.

The remainder of the book concerns itself with free will and the problem of evil (see "The Book of Job," Chapter 6). Schroeder uses the fairly standard kabbalistic argument that God withdrew in order to give us free will. Schroeder's answer to the question, "How can we have free will if God knows the future?" is clever but unsupported in any way: God lives in Eternity; Eternity is timeless, so everything happens at once from that vantage point; and the photon, for which time does not exist, is the link between time and Eternity. Why the photon? No answer; again, he seems to think that stating it is enough.

Using the Bible to "prove" or "verify" modern science is a dangerous game, and two can play it. For example, 1 Kings 7:23 says, "And he made a molten sea, ten cubits from the one brim to the other: it was round all about,... and a line of thirty cubits did compass it round about," with almost identical wording in 2 Chronicles 4:2. Was pi equal to 3 in the days of Solomon? Or can we say that the people in those days noticed a certain regularity between the diameter and the circumference of a circle but did not have the tools to understand this regularity very well? If that is so, why can we not say the same about the first chapter of Genesis?

Similarly, I have found in the Hebrew Bible about a dozen references concerning the Heaven above and the earth below, the ends of the earth, the corners of the earth, and the length of the earth. The most telling is perhaps Daniel 4:20, "The tree that thou sawest, which grew, and was strong, whose height reached unto the Heaven, and the sight thereof [reached] to all the earth." Can you build a tower so tall that it can be seen from anywhere on earth? Not unless the earth is flat.

I am not so foolish as to believe that such inaccuracies "prove" that the Bible is wholly wrong or even untrustworthy. A great deal of modern scholarship and archeology have established the rough accuracy of, let us say, those parts of the Hebrew Bible that postdate David or perhaps Deborah. Many of the earlier portions may be accurate in outline as well. Other parts, such as the Books of Esther, Jonah, or Job, are generally

125

accepted to be fictions designed to make a point. Yet other parts, such as the Song of Songs, are poetry. And other parts, such as the first chapter of Genesis, are allegory. None of these observations diminishes the Bible as a historical or religious document. What diminishes the Bible is the insistence, against all odds, that every word in it is literally true and that God did not have the imagination to write allegorically.

The Hebrew Bible

If the Bible is absolutely true, it ought at least to be consistent. Yet it was a study of the inconsistencies in the Hebrew Bible that led a great many scholars to conclude that it most likely had a multiplicity of authors. Indeed, you do not have to know much Hebrew to realize that the Bible is chock full of typographical errors.

The text we use today is called the *Masoretic text*, after a group of Jewish scholars called the *Masoretes*. The word is related to the Hebrew words for *handing down* and *tradition*. Working between about the years 700 and 900 C.E., the Masoretes were faced with a multiplicity of Torah scrolls, each a little different from the others. In addition, now that Hebrew was no longer the vernacular, people in general did not know how to pronounce all of the words. This is so because Hebrew is written without vowels, srt f lk ths [sort of like this]. If you can interpolate the vowels correctly, you can read the Hebrew. However, if you interpolate them incorrectly, you can often get a different meaning. Chaim Potok [1982] points to one of the most famous passages in the King James Version:

> Yea, though I walk through the valley of the shadow
> of death, I will fear no evil: for thou art with me
> (Psalms 23:4).

Potok, a novelist with rabbinic training, was the secretary of the committee that prepared the Jewish Publication Society's

translation of the Hebrew Bible. He explains that the King
James translation is based on the Septuagint, a translation of the
Hebrew Bible into Greek. That is, the King James translation is
not based on the original Hebrew but on a Greek translation of
the Hebrew text. When the Septuagint was prepared, a certain
word *tsalmut*, which means *deepest darkness*, had become
obsolete in Hebrew, so it was misread as *tsel mavet, shadow of
death*. The same verse in the JPS translation is rendered

> Though I walk through a valley of deepest darkness,
> I fear no harm, for You are with me.

Out of respect for tradition, the committee agreed to a
footnote acknowledging, "Others 'the valley of the shadow of
death'." Note also the substitution of *harm* in place of *evil*.

The Masoretes, then, had to choose the "correct" texts as
well as decide the "correct" pronunciations. For the
pronunciations, they developed a series of diacritical marks that
indicated the vowels. Even so, the few early texts we have today
are not identical, and the modern version of the Masoretic text
dates only to the invention of the printing press. There are also
differences between the Masoretic text and the Dead Sea scrolls.
Thus, although most of the Bibles printed today and every Torah
scroll in every synagogue are said to be copies of "the"
Masoretic text, there is in reality no unique Masoretic text but
rather an agreed-upon standard. How do we know that it is the
"correct" text?

* * *

One of the first things you learn when you learn to read the
Torah is that some words are not pronounced as they are written.
Specifically, the pronoun *she* is written היא, sometimes
transliterated HY'. The pronoun *he* is written הוא, or HV'. The
only difference between the two words is the height of the
middle letter, and evidently some Masorete got carried away

from time to time and wrote הוא when he meant היא, probably because הוא is the more common. This error, clearly no more than a typographical error, is reproduced faithfully in every Torah scroll and every Hebrew Bible printed today. Such "accuracy" does not give me confidence in the historical accuracy of the Masoretic text, let alone confidence that it is the absolute word of God.

The Bible has more-important inconsistencies. For example, the books of Kings and Samuel tell much the same history as Chronicles, but with slight differences. Robin Lane Fox [1991], a Reader in Ancient History at New College, Oxford, cites an early Greek translation of Samuel, which gives different chronologies from the Masoretic version. Lane Fox thinks the Greek version is more nearly correct. Presumably, the Hebrew text from which this Greek translation was derived has been lost. Lane Fox further speculates that Chronicles was based on an earlier and more accurate version of Samuel itself. He states flatly that the version that is read today in churches and synagogues is "an arbitrary version, padded out with later material."

The Dead Sea Scrolls likewise contain manuscripts similar to the Masoretic text but not identical to it. Lane Fox cites a text of Jeremiah that is one-eighth shorter than ours, a text of Samuel that is "not so very close to" the Masoretic text, variants in a text of Job, what he calls a "doctored" text of Ecclesiastes, and two different texts of Isaiah. The five books of Moses, the Torah, show somewhat fewer differences, but, according to Lane Fox, the Dead Sea Scrolls also contain unfamiliar bits of Deuteronomy and an unknown text of Leviticus, for example. Lane Fox stresses that these textual differences are not "silly mistakes"—nor what I have been calling "typographical errors"—but rather are "independent streams of development." [See also VanderKam, 1994]

Let us concentrate for a moment on the Book of Genesis. Many of the stories in that book are told twice, sometimes three times, with slight differences. In one version of each story, God is usually called *YHVH*. This is the root from which "Jehovah" derives, but it is traditionally pronounced "Adonai" in order to avoid the possibility of desecrating it. In the other version of the story, God is called "Elohim." Adonai is usually translated Lord, whereas Elohim is translated God. (Elohim is itself an interesting word for monotheists, incidentally, since it is the plural of Eloah, one of the Hebrew words for God. The Bible uses the plural form Elohim, usually, but not always, with a singular verb; for example, "And God said, Let *us* make man in *our* image" (Genesis 1:26) uses a plural verb, whereas "And the Lord God said, Behold, the man is become as one of us" (3:22) uses a singular verb. Using the plural verb and the plural ending with reference to God suggests at a minimum that the Hebrews were not wholly monotheistic at the time that the Bible was being composed.)

On the basis of such evidence, Bible scholars have deduced that the Hebrew Bible is composed of at least four documents that were amalgamated by an unknown person or group known as the *redactor*. Biblical literalists reject this *documentary hypothesis* out of hand. The evidence, however, is so strong that the Anglican (Episcopal) bishop John Shelby Spong [1991, p. 43] says flatly that the documentary hypothesis is almost incontrovertible. He calls it a tragedy that the typical worshipper has "not been introduced to these insights." In a good introduction to "these insights," Spong outlines how someone, probably during the reign of Solomon in about 950 B.C.E., wrote down the oral tradition of the history of Israel. Because this author used *YHVH* for God, his document is called the J document (J not Y because the original work on the documentary hypothesis was published in German and *YHVH* is rendered *JHWH* in German). Spong notes, for example, that the J document's version of the Ten Commandments in Exodus 34 does not include the commandment to rest on the Sabbath because the seven-day creation story was a later development and had not yet been written; this is the kind of evidence scholars

129

use to deduce the provenance of the J document and separate it from the others. [See also Anderson, 1966; Lane Fox, 1991]

The Jewish nation was split immediately after the death of Solomon, and those in the Northern Kingdom went into competition with those in the Southern Kingdom. They wrote their own version of the oral tradition and for unknown reasons called God Elohim, not *YHVH*. Their version became the E document. Still later, in about 621 B.C.E., during the life of Jeremiah, the priests discovered or claimed to discover a lost book of the law. This book eventually became Deuteronomy. Finally, after the fall of the first Temple in 586 B.C.E. and the Babylonian Exile, the Priests developed laws such as the laws of Sabbath observance. They doctored the earlier texts and added some of their own, most particularly the seven-day creation story, which they then used to justify Sabbath observance. The first chapters of Genesis are part of the Priestly, or P, document.

Richard Elliott Friedman is a professor at the University of California, San Diego, and a former student of the influential Biblical scholar Frank Moore Cross at Harvard. In *Who Wrote the Bible?* Friedman [1987] reviews the evidence in favor of the documentary hypothesis and notes not only that the J and E documents have their own names for God, but also that each has its own language and its own concerns. Based on such considerations, Friedman further concludes that the portions that use Elohim for God may derive from either the E document or the P document; that is, the Priests added to or edited the E document.

Let me give one example. How many animals did Noah take aboard the ark? Two of each kind, you say? That is not exactly what the Book of Genesis says. Rather, the story begins with Elohim telling Noah (Gen 6:19, 20),

> And of all that lives, of all flesh, you shall take two into the ark to keep alive with you; they shall be male and female [note the terms, male and female]. From birds of every kind, cattle of every kind, every

creeping thing on earth, two of each shall come to you to stay alive.

Friedman assigns this passage to the P document. Here, incidentally, I have used the Jewish Publication Society's 1962 translation because it represents a monumental effort by a team of scholars to prepare an accurate translation in modern English. It is therefore closer in meaning to the original Hebrew than the more traditional translations. [Potok, 1982]

In the very next chapter (7:2, 3), it is Adonai, not Elohim, who tells Noah,

Of every *clean* [italics mine] animal you shall take seven pairs, males and their mates [note the terms, males and their mates], and of every animal which is not clean, two, a male and its mate; of the birds of the sky also, seven pairs, male and female, to keep seed alive upon the earth.

Friedman assigns this passage to the J document. Later in the same chapter, we learn that (7:8, 9)

Of the clean animals [italics mine], of the animals that are not clean, of the birds, of everything that creeps on the ground, two of each, male and female, came to Noah into the ark, as God [Elohim] had commanded Noah.

Friedman assigns this passage to the P document. Then, finally (7:14, 16),

... they and all beasts of every kind, all cattle of every kind, all creatures of every kind that creep on the earth, and all birds of every kind, every winged thing [went into the ark]... two of all flesh in which there was breath of life. Thus they that entered

131

comprised male and female of all flesh, as God [Elohim] had commanded him.

Friedman assigns this passage also to the P document.

So what do we have? First, Elohim tells Noah to take *two* of each kind into the ark. Then, Adonai says take *seven* pairs of each "clean" animal. Next, the P document tells us, almost as if P had been eavesdropping on J, that no, Noah has taken *two* of each kind, whether clean or not, as *Elohim* has commanded him. Finally, the P document repeats that Noah has taken *two* of all flesh into the ark. In short, there are two contradictory statements: Noah took two of each kind into the ark, and Noah took seven.

What about the term "clean"? It is taken to mean fit for sacrifice. Domestic animals are fit for sacrifice, whereas predators and animals that have wounds of any kind are not. The stricture against wounded animals means, in effect, that animals that have been hunted or trapped are necessarily unclean. The traditional explanation of these passages is that Noah was instructed to bring aboard seven pairs of each kind that was fit for sacrifice and two of all others. Presumably he did so in order to ensure that there would be enough clean animals to sacrifice (Spong thinks, therefore, that this passage was altered by the Priests). Why then does 7:3 say, "of the birds of the sky also, *seven* pairs"? The birds of the sky, whether predators or not, will almost certainly have to be shot or trapped and will therefore be wounded if they are available for sacrifice. That is, the birds of the sky can never be fit for sacrifice, so one pair would have sufficed.

Other problems with the Masoretic text involve dialect. Apparently, the Bible as it has been handed to us was written in a dialect of Hebrew that did not survive. In any event, many words did not survive, and their meanings are not known. The JPS translation is peppered with footnotes that read, "Meaning of Heb. uncertain." One of the more well known passages of Genesis is, "And the earth was without form, and void" (1:2).

"Without form and void" or "formless and void" is the standard translation of the Hebrew "tohu va-vohu," but no one really knows what these words mean. Robert Graves and Rafael Patai [1983] think that they derive from the words for a sea monster and a land monster, but that these original meanings may have been lost.

The Book of Jonah

The Book of Jonah ought to be exhibit 1 in the case that the Bible is not literally true; it is a short story and nothing more. Jews traditionally study the Book of Jonah on Yom Kippur, the Day of Atonement, evidently because the book is about sinning and repenting. I have therefore read it many times, both in Hebrew and in English, and discussed it in groups led by rabbis several times on Yom Kippur. It is a fascinating story and about much more than repentance. It contains many truths (though not many facts), but it is not a true story in the sense of being historically accurate.

In the period after the Babylonian exile, authors often did not claim credit for their works but rather credited them to earlier authorities in order to give them higher status and make them more persuasive. Thus, we can infer that their works were intended as polemics, that is, controversial works designed to sway the audiences to their authors' points of view. Jonah ben [son of] Amittai was an obscure prophet mentioned only once, in 2 Kings 14:25, when his name was evidently adopted by the author of the Book of Jonah. Lane Fox dates the Book of Jonah to the fourth century B.C.E. We have no idea who wrote it.

Isaac Asimov, [1981] the well known science fiction writer and popularizer of science, has also written a study of the Bible as a historical document. He shows clearly that the Book of Jonah is an anachronism. The real Jonah lived during the reign of Jeroboam II, one of the kings of Israel. We therefore can infer that Jonah was active as a prophet about 780 B.C.E. The Book of Jonah refers to Nineveh as a great city, capital of the Assyrian

133

Empire. Nineveh, however, did not become the capital of the Assyrian Empire until about 100 years after the reign of Jeroboam, according to Asimov. It remained the capital for only about 75 years. It was during those 75 years, however, that Assyria conquered Judah. In consequence, Nineveh was widely known, and the author of the Book of Jonah must have incorrectly assumed that it had been a great city during the time of the real Jonah.

Jonah is remembered primarily for having been swallowed intact by a big fish (not necessarily a whale, incidentally, though a fish to the Biblical chroniclers probably included any vertebrate that lived in the sea). He was later regurgitated in good condition on the beach. Beside the creation myths, this is possibly the best-known part of the Hebrew Bible. It is the least important part of the story but also the hardest to believe. Indeed, it seems to me that you have to be particularly credulous to accept the story of Jonah at face value and not recognize it as either a fiction or, at the very least, an exaggeration that simply cannot be taken literally.

Let us look at the story of Jonah in more detail. God tells Jonah to go to Nineveh, the capital of the Assyrian empire, and announce that Nineveh will be destroyed unless its people repent. Assyria had oppressed Israel cruelly, so Jonah is naturally afraid and flees to Tarshish, which is thought to lie on the coast of what is now Spain, that is, in the opposite direction from Nineveh. A great storm arises, and even the mariners are afraid. The men pray, each to his own god, but Jonah goes down into the hold and falls asleep. It is hard to believe that he could fall asleep in such a storm, but a common interpretation is that he was depressed. At any rate, the ship's master wakes Jonah and demands that Jonah pray to Jonah's god (as opposed to God or Adonai; the ship's master does not know the identity of Jonah's god) in hope that the god will listen and the storm will abate.

The god does not listen, so the men cast lots as a means of divining the cause of the storm. The lots single out Jonah. Jonah says that he is a Hebrew and worships Adonai, the God of

Heaven, who made the sea and the land. He tells the men to throw him overboard in order to save the ship. The men resist but eventually toss Jonah overboard and make a sacrifice to Adonai (not to their own gods).

Adonai prepares a big fish to swallow Jonah, and Jonah offers a prayer of repentance. He admits that he thought he had got out of the sight of Adonai and promises to fulfill his vows. In literal translation, the fish vomits him onto the land.

Jonah goes to Nineveh, which is an enormous city, three days' walk from one end to another. He prophesies that, unless the Assyrians repent, the city will be destroyed in forty days. The Assyrians repent, and Adonai spares the city.

Jonah gets angry that the prophecy is not carried out. He wishes for death. Adonai provides a plant (no one is certain what kind) to provide some shade for Jonah, but then he sends a worm to kill the plant. Jonah grieves for the plant and again asks for death. Adonai answers that Jonah has no business grieving for the plant, which he did not work for, and at the same time argue that Adonai should not care for the inhabitants of Nineveh.

The story is frankly impossible to believe. Besides the fish story and the exaggeration of the physical size of Nineveh, we have the unlikely behavior of the men aboard the ship. Jonah, fleeing from God, goes on board a ship, gets caught in a storm so powerful that experienced mariners are afraid—and goes below decks and falls asleep. The men wake him up, grill him, refuse to throw him overboard—and then give in and sacrifice not to their own gods but to Adonai. Jonah goes to Nineveh, where they speak Assyrian, a semitic language related to but different from Hebrew, and actually succeeds in convincing the people there to repent. The story is extremely far-fetched.

The Book of Jonah is a fiction, most probably designed to counter the belief that each nation has its own god and that you can get away from God (that is, Adonai) by going elsewhere, outside God's domain. The moral of the story of Jonah is that God rules everywhere, and you cannot escape God by any artifice. The sailors, who know that Jonah has fled from Adonai,

135

appreciate this fact and, apparently, so do the inhabitants of Nineveh.

* * *

The story of Jonah is assumed to have been written after the exile, that is, after the return from Babylon. After the destruction of the Temple in 586 B.C.E., many Jews thought, consistently with the beliefs of the time, that they lost the Temple because the gods of the other nations were stronger than their God, not because of any failing in themselves. Pagan and polytheistic beliefs obviously had more than a foothold among the people of Israel; after all, those beliefs are what the prophets fulminate against most often.

Many of us think that monotheism came to Israel fully formed when God spoke to Abraham or perhaps to Moses. That is simply not so. In a very influential book, *Judaism as a Civilization*, the rabbi Mordecai Kaplan [1967, Chap. 25] describes how Jewish beliefs evolved gradually from polytheism through *henotheism*, or the belief in national gods, to monotheism. The debate between monotheism and henotheism and the debate between universalism and nationalism were raging in full force when the Book of Jonah was written. Jonah is a short story designed at least in part to examine this question and comes down on the side of monotheism and universalism. The author used the name of Jonah to give his polemic an authority it would not otherwise have had. The book is a story, not a carefully reasoned argument, and it would not have been credible unless it had been ascribed to a prophet.

* * *

The last chapter of Jonah, the story about the plant, is about compassion and is not connected to the arguments about monotheism or universalism. Here, the author seems to be arguing that God is compassionate, despite his earlier threats to destroy Nineveh for its evil ways. God seems pleased that he has averted disaster, because, after all, there are a lot of people in Nineveh who cannot tell their right hands from their left hands (that is, children, who are presumed innocent), as well as many beasts (who are also innocent). In short, God was willing to destroy the entire city, children and cows and all, if the wicked inhabitants did not repent. This takes me straight back to Kushner's argument that he could fear such a God but not worship him. You could, I suppose, argue that God knew all along how the story would come out and therefore knew that he was not going to destroy Nineveh, but there is not the slightest hint in the story that the author intended that interpretation. Besides, I do not think that the Biblical chroniclers intended to present God as a cosmic bluffer; the God of Jonah would have destroyed the innocent along with the guilty if Nineveh had not repented.

The Book of Jonah is not the only supposedly historical book or section that is demonstrably fiction. The Book of Esther, which is set in historical times, is also almost certainly not historically accurate. Anderson identifies the king in Esther with the Persian king Xerxes I (ca. 519-465 B.C.E.). It is hard to believe that the Jews could have rebelled, with the permission of Xerxes, all over Persia, and yet there would not be a single record of the rebellion in any other known document. It is possible that Esther is an exaggeration based on a grain of truth, but I think it more likely that it is a novel written by someone in Persia. Unlike the characters in the Book of Genesis, the characters in the Book of Esther are unflawed and one-dimensional. Lane Fox dates the Book of Esther between 280 and 180 B.C.E., or well after the reign of Xerxes.

The Gospels

J. P. Moreland is a philosophy professor at Biola University in California and works with the Campus Crusade for Christ. Kai Nielsen is an atheist philosopher and Professor Emeritus at the University of Calgary. In one weekend in 1988, Moreland and Nielsen made a series of speeches (not really a debate) and compiled them as a book. [Moreland and Nielsen, 1993] They invited five other philosophers to comment on the speeches and, further, added rebuttals of their own.

Nielsen applies a linguistic analysis and tries to show that a belief in the traditional Western God is irrational because the concept is incoherent. For example, Nielsen asks what it means to say that God is both immanent in the world and transcendent to it. I am unimpressed by such arguments and see no reason why I cannot stand waist deep in a swimming pool and therefore be both within the water and simultaneously above it. At any rate, Nielsen attacks a straw man. You can render his arguments irrelevant simply by defining God as a purposeful creator and omitting the properties that he considers incoherent, as I will do below.

Moreland presents the case not only that theism is rational but further that Christianity is true whereas all other religions are false. His case for the second claim is a circumstantial case that the Gospels are accurate because they are first-hand accounts of the Resurrection, whereas the sacred books of other religions are mythology in the sense that they were passed down orally for several generations before being written. Implicit in Moreland's argument are the assumptions that oral history is necessarily inaccurate and that written history is impervious to distortion. Neither assumption can stand scrutiny; we have already seen that the written parts of the Hebrew Bible contain errors and suggested that the orally transmitted parts contain truths. We have no reason to accord more integrity to the Gospels. (The Oxford philosopher Richard Swinburne [1996] adds that the founding myths of Judaism and Islam, such as the parting of the Red Sea, are either not claimed to be miracles or can be explained as natural phenomena; I will show below that the Jesus myths can also be explained by natural phenomena.)

Moreland argues that the Gospels are contemporaneous accounts that accurately portray the events they describe. He cites archeological evidence that suggests that the description and the location of the tomb of Jesus are accurate. He notes that the tombs of other holy men were venerated, whereas the tomb of Jesus was not. He notes further that women, who were not considered credible witnesses, discovered the empty tomb; a fabricated account would have used men, since they would have been considered more credible as witnesses. Finally and most importantly, he argues that people saw Jesus after his death by crucifixion. If that is true and if Jesus was truly resurrected, then I must agree with Moreland that the case for theism is proved, whether or not Jesus is God or the Christian interpretation is accurate in detail. I cannot accept Nielsen's argument that firm evidence of a resurrection would be no more than an interesting physical fact, no different from a "miraculous" recovery from a usually fatal disease. The physical changes that take place in a dead body in three days ought to refute that line of thought very effectively, but Nielsen seems as set in his belief as a scriptural literalist.

The Gospels are in fact not contemporary accounts, and Moreland admits as much; they may have been written as late as 70-100 C.E., though Moreland cites a source that argues for 40 C.E. for the earliest Gospel. The Crucifixion is thought to have occurred in 33 C.E. In addition, the Gospels are not independent accounts. [Kee et al., 1965] By analyzing the texts and comparing them, New Testament scholars have concluded that the Gospels of Luke and Matthew are very probably based on Mark and a hypothetical lost source called the Q document. The Gospel of John presents a special problem: It seems to have been written much later and contains differences in chronology and tone from the other three. It may be based on the first three Gospels or on oral traditions concerning the same topics. In consequence, there are probably not four independent accounts, but two, one of which, the Q document, has not survived. Thus, interpreting the Gospels literally presents the same problems as interpreting the Hebrew Bible literally.

139

Moreland's argument that the tomb was not venerated is not real evidence for the Resurrection; perhaps it was not venerated because the disciples had stolen the body and the tomb was empty or because Jesus was thrown into a mass grave and there was no tomb. Likewise, his argument that women were not considered credible witnesses does not hold water. Even if women were not acceptable as witnesses in the legal sense, it is clear that the people in the Gospels believed the women; for example, the Samaritans "believed in him because of the women's testimony" (John 39:1; Revised Standard Version).

* * *

Moreland nevertheless makes a good circumstantial case; the Gospels clearly contain a great deal of historical material. What else do they say about the Resurrection?

John (chap. 20) tells of Mary Magdalene's going to the tomb of Jesus and finding it empty. She speaks to a man there but thinks it is the gardener (20:15)—that is, she does not recognize him as Jesus. The man says "Mary," and she suddenly recognizes him. She tells the disciples, "I have seen the Lord." Later, Jesus appears to the disciples and shows them his hands and his side. (20:20) Only then are they described as "glad" to see the Lord; that is, they do not recognize him either. A few days later, Jesus appears to Thomas, (20:26-29) who also will not believe it is Jesus unless he can see and feel the wounds. Finally, Jesus appears to the disciples when they are fishing, and they do not recognize him.

Luke tells of two men who are going to Emmaus, a village near Jerusalem. They see Jesus, but "their eyes were kept from recognizing him" (24:13). They speak with Jesus, tell him of the empty tomb, invite him home for bread. Jesus speaks theology with them. Then, (24:31) they recognize him, and he vanishes.

In none of these encounters do the people recognize Jesus. You can explain these incidents away if you like. Of course, they would be skeptical of seeing a dead man and demand proof.

But the undercurrent that Jesus appears and is not recognized, even after he is known to be prowling, is striking and at a minimum casts serious doubt on Moreland's argument. It is at least plausible that they met someone else and later convinced themselves that it had been Jesus.

Matthew relates that Pilate sets a guard over the tomb so that people will not be able to say that Jesus has risen from the dead. (Matthew 27:62-66) After the tomb is discovered to be empty, the priests bribe some soldiers to tell people that Jesus's disciples came and carried him away while the soldiers were asleep; the Jews tell this story "to this day." (28:11-16) It seems unlikely in the extreme that the soldiers would have dared to admit that they were asleep on the job; "to this day," you can be severely punished for falling asleep while on guard duty. Therefore, we can reasonably infer that this is a defensive story, almost certainly not true, but designed to counter the widespread belief (knowledge?) that the disciples had stolen the body. Further, Mark ends with Mary Magdalene and some others coming upon a young man in the opened tomb. If it is the same young man whom Mary mistakes for the gardener in John, Mark gives not the slightest hint that he is in reality Jesus.

* * *

Hugh Schonfield was a New Testament scholar who was educated at King's College, London, and earned a doctorate in sacred literature at the University of Glasgow. He was a prolific author but received attention mostly for *The Passover Plot*, [1966] in which he argues that Jesus and his followers planned the Crucifixion but Jesus was unexpectedly killed by a spear. Regarding the Resurrection scenes, Schonfield notes that the first person to see Jesus is Mary Magdalene. He argues that she is possibly somewhat unbalanced, since Jesus has earlier had to cast seven demons out of her. Her eyes are clouded with tears. She is devoted to Jesus, in love with him, and stricken with grief. Possibly she is deluded into believing that Jesus is in love with

141

her. At any rate, she runs to tell the disciples that she has seen Jesus.

Schonfield believes it plausible that the myth of the Resurrection began with the delusion that Mary had seen Jesus. His interpretation of Luke's text is that the stranger departs, as opposed to vanishes, and the disciples later talk themselves into believing it was Jesus: "Did not our hearts burn within us while he talked to us on the road, while he opened to the scriptures?" (Luke 24:32) Emmaus is only a few kilometers from Jerusalem, and Schonfield speculates that the man in the tomb was the same as the man whom the disciples saw; in both cases he was merely carrying a message from Jesus but was later assumed to have been Jesus.

Schonfield's main thesis, by the way, is that Jesus and his followers staged the Crucifixion in order to ensure that the prophecies would be carried out. This idea might be somewhat fanciful, but it is based on a close reading of the Gospels and almost nothing else. That is, Schonfield's argument is no different in that respect from Moreland's, yet he comes to a very different conclusion.

* * *

The Anglican bishop John Shelby Spong [1991] was raised as a Biblical literalist but, in his terms, rethought the meaning of scripture and no longer believes in the inerrancy of the Bible. As he puts it, his view is the view of a practicing Christian, not a hostile critic. He writes with the intention of "rescuing" Christianity, because he believes that "theological truth" must be "separated from pre-scientific understandings and rethought in ways consistent with our [present day] understanding of reality." His view of the Resurrection is substantially more radical than that of Schonfield: Spong believes that the Resurrection never truly happened.

Moreland and Schonfield agree that the events of the Gospels are real events. Spong does not. Instead, he argues that

142

the Gospels are embellishments in the tradition of Jewish *midrash*. Midrash (plural *midrashim*) is a method of interpreting the Bible by telling a story; many of the parables of Jesus are midrashim. The rabbis of the first century commonly used midrashim to illuminate a passage of the Bible or to tell a story with a moral. Spong thinks that the Resurrection stories of the Gospels are such stories. He notes, for example, that Jews believed that Elijah would return to earth to presage the coming of the Messiah. Elijah was supposed to have been taken bodily into Heaven in a fiery chariot. Similarly, Moses died, but no one knows the place of his burial. Spong considers it but a short step to write similar midrashim with Jesus in place of the original hero: Jesus died and no one knows the place of his burial; Jesus was taken bodily to Heaven. A critic of Spong, however, would note in rebuttal that Jewish midrashim rarely, if ever, retold an old story with a new hero.

Spong argues, nevertheless, that the Gospels are not written in what he calls "linear time"; that is, they are not written chronologically but are simply a collection of midrashim. Thus, Spong views Luke's story about the boy Jesus being lost for three days in Jerusalem as a literary device designed to anticipate the three days between the Crucifixion and the Resurrection. It makes no sense, says Spong, for Jesus to call Simon the rock upon which the church will be built, because the disciples at that time have no concept of an institution called a church. These are anachronisms inserted by later writers or redactors with the gift of hindsight.

Spong notes that the Epistles of Paul were written before any of the Gospels. Paul states flatly that Jesus was buried. He says (1 Corinthians 15:3-8) that Jesus was resurrected and appeared to Peter and the twelve and the five hundred and the apostles, but Spong sees this passage as a standardized liturgical passage or a dogma that Paul was simply repeating. For Spong, this passage gives no indication of a *bodily* resurrection but rather means that the five hundred or so people received "revelatory visions" of Jesus at the right hand of God. We now interpret it as meaning

143

bodily resurrection because of the distorting lens of the Gospels, says Spong. But Paul makes no distinction between the Resurrection and the Ascension, that is, between the appearance of the crucified Jesus on earth and his subsequent ascent into Heaven. Paul believed that God had raised Jesus directly to Heaven, but Spong argues that Paul did not mean a bodily ascension to Heaven: "Flesh and blood cannot inherit the kingdom of God, nor does the perishable inherit the imperishable," says Paul (1 Corinthians 15:50). Spong regards this verse as a specific statement that bodily resurrection is impossible.

The first Gospel, chronologically, was that of Mark. The later Gospels, as I have noted, were based partly on Mark's. Spong calls the Gospel of Mark "Christian Midrash at its best." He argues that Mark had very little factual material: Jesus came from Galilee, he had something to do with John the Baptist, he told parables, he journeyed to Jerusalem, he was crucified, and his disciples had what Spong calls a powerful experience that made them think that God had raised him from the dead. To fill in the blanks, the later followers of Jesus searched the Hebrew Bible for material that could validate their claims. Thus, Spong sees the Gospel of Mark as retellings of the ancient Bible stories with Jesus in place of Abraham, Isaac, Moses, Joshua, Samuel, David, Solomon, and so on. Mark's Gospel ends with not a word about the Resurrection.

The next Gospel, the Gospel of Matthew, embellishes on Mark's account, according to Spong. For example, a young man clad in white becomes, for Matthew, an angel. In almost every instance where Matthew rewrote Mark, Spong has found evidence for a midrashic rewriting. The cave, the guards, and the stone, Spong says, were borrowed from a story about Joshua, who similarly imprisoned five captured kings in a cave. In Matthew, Jesus becomes a "semiphysical being" who talks to the women.

Luke was either a Gentile or a thoroughly assimilated Jew. He wrote the Gospel of Luke perhaps 20 years after Mark had written his. Spong sees Luke's account as a version that would appeal to Gentiles because it shows a pious or heroic person who is taken directly up to Heaven; this was an image that Gentiles could readily understand because it was common in mythology. At the time of the Gospels, many Gentiles were drawn toward Judaism, perhaps through intermarriage, but did not formally convert and did not follow the dietary laws; they are sometimes called the God-fearers. Possibly Luke directed his writings toward them.

Like Matthew, Luke embellishes Mark's story. In particular, Luke's is the only Gospel that tells the story of the two men who meet Jesus on the road to Emmaus; this story more than any other emphasizes that Jesus was resurrected bodily and walked the earth after his death. Where Mark has a young man dressed in white meet Mary Magdalene at the tomb, Matthew has an angel, and Luke has two angels. Furthermore, the women give the message to the disciples, who are in Jerusalem in Luke's version. Peter visits the tomb and finds it empty except for the grave clothes. The tomb, according to Spong, has now, in Luke, become not the sign of the Resurrection but the proof. Finally, the Book of Acts, which is thought to have been written by Luke, depicts the resurrected Jesus ascending physically to Heaven as two angels make another appearance.

The first three Gospels are called the *synoptic Gospels* because they share a similar viewpoint. The last Gospel, the Gospel of John, is apparently drawn from a different tradition and presents somewhat different characters, different chronology, and different locations. John, according to Spong, contains relatively late and highly developed theological material, as well as more-primitive descriptions of events. In John, Resurrection and Ascension are not so clearly distinguished, but, after his resurrection, Jesus is at least half spirit in that he can walk through walls. The earliest accounts do not call Jesus God, but John insists on it and has Jesus repeatedly use words reminiscent of the words God used to Moses: "I am."

145

Spong flatly denies the stories that surround the Resurrection. Instead, he reconstructs the events in this way: Jesus is arrested and crucified. His body is thrown into a common grave. Mary Magdalene possibly tries to locate his body but is unsuccessful. Simon (later called Peter) denies Jesus and flees, perhaps to Mary Magdalene's house in Bethany. As soon as it is safe, he returns to his home in Galilee and takes up his trade, fishing. As a fisherman, Simon has plenty of time to think, and the effect of Jesus on Simon has been enormous. Eventually, Spong speculates, Simon decides that what he experienced with Jesus has been real. The Crucifixion was not a punishment but the culmination of a drama acted out in the real world. Simon and his colleagues return to Jerusalem, and Spong presents a circumstantial case that the return of Simon and his band forms the basis for the (apochryphal) story of the journey of Jesus and his band to Jerusalem on Passover.

For Spong, the Resurrection was a real event but not a physical event. It did not happen in Jerusalem but rather in Galilee when Simon realized that God's love was unconditional. He mentions the possibility that Simon was "delusional" but dismisses the idea with no further discussion. John Polkinghorne [1994] similarly believes that hallucinations could not have been the "enduring basis" of early Christianity but likewise offers no evidence to the contrary.

* * *

My own pet theory, which I will not bore you with, is that Jesus was released or escaped, and everyone knew it. Hence, his disciples made up a cover story about the release of a thief. Somewhat unimaginatively, they called the thief Barabbas or, in some manuscripts, Jesus Barabbas. [Kee, Young, and Froehlich, 1965] Barabbas is a Greek adaptation of the Hebrew Bar-Abba, son of Abba. Abba is a name, but it also means father. I maintain that there can be only one Jesus Son of Father in the New Testament.

146

The English classicists and writers on Paganism, Timothy Freke and Peter Gandy, [2000] take the extreme position that Jesus never existed. They note strikingly precise parallels between the Jesus myth and the myths of Dionysus, Osiris, Mithras, Adonis, and other gods. According to Freke and Gandy, the myths of these national gods were identical, and everyone knew that they were myths. Indeed, they show that Christians were sometimes ridiculed for pretending that their god, Jesus, was real.

* * *

I have presented three interpretations of the same books, as well as vignettes of two other interpretations. They come to remarkably different conclusions. It is only a small exaggeration to say that you can read into the Bible almost anything you like.

I began this section with Moreland's argument. I have used the two other accounts to help examine the validity of Moreland's argument. It now seems safe to say that Moreland is guilty of selective use of evidence. He quotes the Gospels as if they were a coherent account and ignores passages or interpretations that cast doubt on his thesis. His case is by no means proved.

The scientific literature

I can almost hear my critics shouting, "Well, you believe literally in what you call 'the literature'; what's the difference between believing in the literature and believing in the Bible?"

A world of difference. First, I do not believe everything I read in the scientific literature; I read the scientific literature as critically as I read the Bible. Consider, for example, the study on sugar that I cited above. Such studies are usually evaluated by statistical methods, and the experimenters usually estimate the probability that their results could have happened by chance. An experiment is considered *statistically significant* if that

147

probability is 5 percent or less. This probability is called the 0.05 level, pronounced *oh five level*. A result that appears significant at the 0.05 level has approximately one chance in 20 of having happened as a result of chance, because $1/20 = 0.05$. That level of statistical significance is considered marginal; a very convincing experiment should have much higher statistical significance than the 0.05 level. (That is why the statistical significance of Witztum's studies on equidistant letter sequences in the Book of Genesis seems so impressive.)

I am very suspicious of a single experiment that weighs in only at the 0.05 level. This is so partly because of what is called *publication bias*: the practice of publishing positive results and failing to publish negative results. Failing to publish results that are not statistically significant is not necessarily unethical, but all it takes is 15 or 20 studies before you will have a good chance of accidentally getting a result that appears significant at the 0.05 level even though it is not. In consequence, I regard isolated studies reported at the 0.05 level as little more than anecdotal evidence: They suggest more studies but are not conclusive.

In addition, you will occasionally see a study that got a negative result (that is, was not significant at the 0.05 level), but the experimenters noticed some other variable that they were not studying but which appeared significant nevertheless. I usually discount such studies entirely unless the significance is very much higher than the 0.05 level, since all you have to do is look at 15 or 20 variables before you will probably find one that, by chance, appears to be significant at the 0.05 level.

Here I am talking about studies in medicine, epidemiology, or sociology. Physics and chemistry are more cut and dried. If a physicist reports the result of an experiment, I generally believe that result because the experiment can be replicated by any other physicist, and the physicist who cheated or made a mistake would soon be found out. This is so because, as far as we know, any collection of, say, rubidium atoms is precisely equivalent to any other collection of the same number of rubidium atoms. An experiment can therefore be replicated almost exactly. The same is not true of humans or laboratory rats. Experiments that involve animals have many more variables than physics experiments, and that is why I consider physics and chemistry in some sense easier than biology and medicine.

Why do I accept the scientific literature? Because most researchers report accurately what they find and interpret their results carefully; when they do not, they are eventually discovered. Furthermore, the scientific literature is not mired in the dogma of the past but rather is constantly being updated; I do not accept statements in the literature if they are known to be untrue. I look to Newton and Galileo for valuable insights into the history and philosophy of science but do not insist that their writings are the last word on a subject. Finally, I can use my critical faculties to accept, reject, or interpret parts of the literature and do not have to swallow it wholesale. Indeed, that is precisely what Friedman, Lane Fox, Spong, and other Biblical researchers are doing when they study the Bible scientifically. To me, assuming that the Bible is literally true and ignoring years of critical scholarship is like assuming that Newton's *Principia* is true in detail and ignoring relativity and quantum mechanics.

Is the Bible true?

Here, I agree wholeheartedly with Harold Kushner [1976]: It all depends on what you mean by "true." The Bible is a combination of history, mythology, and poetry. Mythology may not be literally true, but a myth may prove enlightening nonetheless. Similarly, poetry: When Shakespeare said,

> Shall I compare thee to a summer's day?
> Thou art more lovely and more temperate

surely he did not mean that your body temperature was more stable than the air temperature on a pleasant morning in June.

149

Lane Fox seems not to accept anything in the Bible as history unless it postdates the stories of David and Saul, though he speculates that the Song of Deborah (Judges 5:2-5) and the Song of Gibeon (Joshua 10:12) are source material, that is, that they date to the times of the events they describe. His view may be a bit extreme, and Lane Fox admits that some Israelites must have come out of Egypt under the guidance of their god Adonai, but he doubts that there was any kind of mass migration of the kind suggested in the Book of Exodus. I agree with him that there is no reason to believe any of the stories of giants on the earth or the Tower of Babel, though there is little doubt that the Babylonians were in the habit of building towers called *ziggurats*, which may have been the archetype for the myth of the Tower of Babel.

As for the stories of Abraham, Isaac, and Jacob, of Noah, of Moses and Joshua, Lane Fox argues that they were first written down hundreds of years after the supposed events, and he doubts the stories entirely. Since I am not a historian, I do not have to be quite so rigid, and I argue that these stories must have a good deal of truth in them. The archeologist Heinrich Schliemann and his successors, after all, located the ruins of ancient Troy by studying Homer, and many considered Homer to be pure mythology, in the sense of being wholly untrue. Likewise, Alex Haley, the author of *Roots*, located his ancestor in Africa and found that his relatives there had an apparently accurate oral history of his ancestor's abduction by slave traders generations before.

Like the Bible, Homer's works were not written down for many generations, but, also like the Bible, they were written to be chanted. This suggests to me that the stories were memorized, rather than paraphrased, and may well have been handed down relatively accurately. For all we know, the padding and other inaccuracies that scholars have detected in the Masoretic text may have been inserted long after the texts were written down, not before.

150

How much truth there is in these stories is another matter. Nor is it always clear that we understand the stories in their context. Consider the novel *A Canticle for Leibowitz*, by Walter M. Miller. [1955] This novel is set in the future, 600 years after an atomic war has apparently destroyed civilization. The Church has returned to its medieval function of preserving "knowledge," though this time the knowledge is often scientific. No one, however, understands the documents that are carefully preserved and illuminated. One, for example, says, "Pound pastrami, can kraut, six bagels, for Emma." Another is a circuit diagram for a transistorized control system. Both are attributed to Isaac Leibowitz, an electrical engineer who sought refuge in a monastery but was later hanged by a mob in the aftermath of the war. Neither is the least bit intelligible to those who preserve and copy them.

The modern world has its own myths. These are sometimes called *urban myths*. Some are obviously ridiculous, like the myth that there are gangs of thugs who will drive around at night with their headlights off. If you flash your lights at them, they will follow you home and murder you. I do not know the origin of this myth, but it is hard to see how they could turn around and get behind you in heavy traffic or, in light traffic, follow you very far without being noticed.

Other myths are more believable. For example, a few years ago, you often heard the claim that in the 1940's the most serious problems in the schools were talking, chewing gum, making noise, in that order, followed by several other discipline problems. In the 1980's, the myth goes, those problems were replaced by drug abuse, alcohol abuse, pregnancy, suicide, rape, robbery, and assault, in that order. No one knew the origin of either list, but the lists were promulgated by the Secretary of Education and were repeated by newspaper columnists, the president of Harvard, the chancellor of the New York City schools, and a nominee for Surgeon General. The problem was that no one seemed to know the source of these lists, according to Barry O'Neill, a management professor writing in *The New*

151

York Times Magazine. [1994] O'Neill set out to find that source. He finally located T. Cullen Davis, a born-again Christian who devised the lists for what O'Neill calls a fundamentalist attack on the public schools.

Davis's 1980's list seems to have been taken from a survey of *crimes*, not discipline problems, that had been sent to principals in the mid-1970's. Davis's list in fact did not begin with drug and alcohol abuse but with rape, robbery, and assault. It was a list of questions, not answers, and began with crimes against people. The order of the questions on the list had no other significance. Davis's 1940's list comes from a teachers' magazine, and O'Neill was unable to verify whether or not it was based on real data. In any case, this list consisted of answers to questions about discipline and had no logical connection to the 1980's list, which presented not answers, but questions, and was not ordered.

O'Neill goes on to trace the lists over a few years and claims to have found close to 250 versions, all, I assume, slightly different. The 1940's list has remained fairly constant, he says, but the 1980's list continues to evolve, and its date is always the recent past.

Another urban myth involves a surgeon who mistakenly amputated the wrong leg of a patient in a Florida hospital. What is left out of this myth is that the wrong leg was itself so severely diseased that surgeons have testified that it would soon have needed to be amputated as well. [*New York Times*, 1995] Thus, although the hospital staff presented the surgeon with the wrong leg, he apparently had ample reason to believe that it was the right leg. In short, the hospital and the surgeon made a grievous error, but the case is not the horror story described by the myth.

These three myths are only a few months or years old. The first is absurd; the second has a grain of truth in it; the third is a half-truth. The last two are in some sense based on fact, yet the facts have been misrepresented as they passed by word of mouth for only a very limited time.

The myths in the Bible were similarly passed down by word of mouth for hundreds of years. The means of transmission, however, was probably more organized in that there was very likely a class of people, priests, who memorized the stories and passed them down from generation to generation. Thus, it is very hard for me to believe that those myths that sound like history are completely false. Surely they have at least a grain of truth in them.

The parts of the Bible that I find impossible to believe are what Lane Fox calls just-so stories. Such stories are found in a great many cultures. They are attempts to explain observed facts. Many Greek myths are just-so stories; so are many of the stories in the Hebrew Bible. For example, let us assume that Isaac and Ishmael were real. That does not mean, for example, that they got their names as the Bible claims, nor that Ishmael was banished into the wilderness, where he ultimately founded a great nation. Rather, the story seems to be a just-so story that "explains" why there were two closely related peoples who shared the same land and possibly the same language but had radically different cultures: The Israelites were agrarian, whereas the Ishmaelites were hunter-gatherers. Indeed, many of the laws in the five books of Moses can be read as instructions to remain agrarian and not to go off and become hunters. For example, the definition of ritually clean as meaning unblemished may be at root a prohibition against hunting and trapping, since these activities almost inevitably result in wounded animals.

Similarly, the myth about the Tower of Babel can be seen as "explaining" the diversity of languages. The myth about Lot's daughters (who bore children by their father) was used to demonize the Israelites' enemies, Ammon and Moab (Genesis 19:37-38), and probably has no basis in fact. And the myth about the Creation was used to "explain" the existence of the world.

Such evidence presents a powerful circumstantial case that, whether or not the Bible was inspired by God, it was handed down in a garbled fashion, with so many inconsistencies that it is

153

often impossible to know what it really means. Hence, I conclude, somewhat equivocally,

> *The Bible is not the literal word of God and is not a history book but is comparatively accurate only when it appears to be relating history.*

Biblical literalists would have us believe that God has seen to it that the right text was handed down to us. If that is so, then how do they account for the inconsistencies and typographical errors? Like astrologers and psychoanalysts, they patch the theory by saying that God must have put those inconsistencies in there for a reason, much as he put the fossils into the ground for a reason. We do not yet know the reason, but it may be revealed to us some day. Thus God carefully programmed the scribes to write *he* when they mean *she* and *she* when they mean *he*. This argument is like reaction formation in psychoanalysis: It is inherently unfalsifiable and therefore has no intellectual validity. It can be applied equally to the Bible as to any other religious document and therefore allows us no way to distinguish between, say, the Bible and competing documents.

Conclusion

Thus, we see that the Bible is not an accurate depiction of the early universe. It is not even an accurate historical document. It is full of inconsistencies, and relatively little of it was written contemporaneously with the events it depicts. The Bible is a valuable and possibly inspired document, but those who insist on its inerrancy gloss over inconsistencies and are frankly unwilling to face facts.

We have now dispensed with the arguments of a great many religious believers, including besides lay people some clergy persons and a few philosophers. Before examining the Arguments of philosophers in detail, let us devote some time to what religious believers find their thorniest problem: the existence of evil.

In sober truth, nearly all the things which men are hanged or imprisoned for doing to one another are nature's everyday performances.... All this nature does with the most supercilious disregard both of mercy and of justice, emptying her shafts upon the best and noblest indifferently with the meanest and worst[.]

JOHN STUART MILL

Chapter 5
The evil that men do

Evil poses a real problem for believers. Prager and Telushkin [1986] go as far as to argue that believers have only one thing to account for: the existence of evil (see also "Wishful thinking," Chapter 3). On the other hand, they say, nonbelievers have to account for the existence of the entire universe. Believers, by contrast, account for the existence of the universe by saying that God created it. Prager and Telushkin's argument is not sound, if only because believers still have to account for the existence of God. That is, if we postulate a God who was powerful enough to create an entire universe, we must ask, Where did such a being come from? If another God created our God, then where did that God come from? The postulate that God created the universe leads to an infinite regression and, in fact, gets us nowhere in terms of explaining anything. It is no more than a substitute for "I don't know," which is by far the better response.

As for the question of evil, it is serious, and religion has a lot to answer for. I have tried to make an accounting in my mind as to whether religion has on balance been a benevolent force or a malevolent force. For every Martin Luther King I can name, I can name a Father Coughlin (the anti-Semitic and xenophobic "radio priest" of the 1930's). For every Gandhi, a Khomeini.

For every Albert Schweitzer (the German philosopher, theologian, and organist who turned to medicine and built and operated a hospital in Africa), a Baruch Goldstein (the Israeli physician who murdered 29 Palestinians in a mosque in Hebron in 1994).

I can tick off a dozen religious wars, from the Muslim conquests to the Crusades, through the Thirty Years War, to the recent civil war in Yugoslavia. In the twentieth century, a million Armenian Christians and six million Jews were murdered, solely because of their religions, by practitioners of other religions. The Middle East is chronically splintered because of religion, as is Ireland, though there may be room for optimism in both arenas.

I consider these truly religious wars. That they have been launched by people who are hungry for power does not mitigate this statement. If Yugoslavia, for example, had not been religiously divided, the war would never have begun. The division of Czechoslovakia, by contrast, was peaceful, at least in part because the Czechs and the Slovaks are virtually all Roman Catholics of closely related but not identical ethnic background. In Yugoslavia, by contrast, atrocities that matched those of the Nazis in quality if not in scale were committed in part because some people found it easy to dehumanize the members of other religions, even though they were almost indistinguishable ethnically and linguistically. Yet I can name very few mass movements in opposition to any of these wars, and very few truly righteous wars. Indeed, I am constantly amazed at how bloodthirsty the Western religions sometimes show themselves to be. (Religious intolerance may well be a peculiar property of monotheism. Carl Sagan [1997] claims that those religions that worship "a supreme god who lives in the sky" are the most "ferocious," as measured by a willingness to torture their enemies, for example. Geoffrey Parrinder [1983] states flatly that intolerance is alien to the character of Buddhism and that Buddhists have typically been hospitable to indigenous religions. I suspect that the same is generally true of polytheistic religions.

As Sagan warns, however, it is too easy to confuse correlation with causality, and I do not claim that monotheism causes intolerance. The apparent correlation, nevertheless, probably bears further investigation.)

People often point, correctly, to the Abolitionist movement of the 1800's as a triumph of religious people over slavery. What is less well known, however, is that the Abolitionists were primarily northerners. Southern ministers so strongly opposed the Abolitionists' views on slavery that two of the largest Protestant denominations, the Methodist Church and the Baptist Church, split into northern and southern branches in 1845. Indeed, specifically to oppose the Abolitionists, the clergy, mainly southern, advanced the argument that Black people were destined to be servants because of the Biblical passages (Genesis 9:21-27) in which Noah gets drunk and falls asleep, naked, in his tent. Noah's son Ham (the presumed ancestor of the Hamites, or Blacks) sees Noah naked, whereas his other sons, Shem (the presumed ancestor of the Semites) and Japheth, cover him. Noah wakes up, realizes what has happened, and pronounces a curse on Canaan, the son of Ham (Genesis 9:24-27):

> And Noah awoke from his wine, and knew what his younger son had done unto him. And he said, Cursed be Canaan; a servant of servants shall he be unto his brethren. And he said, Blessed be the Lord God of Shem; and Canaan shall be his servant. God shall enlarge Japheth, and he shall dwell in the tents of Shem; and Canaan shall be his servant.

To my mind, Noah is at fault for getting drunk—not Ham, and certainly not Canaan. Indeed, it is not at all clear why Noah curses Canaan, who did not see Noah naked, rather than Ham, who did. Nevertheless, religious apologists for slavery and segregation have pointed to this passage as justification for their stands on these issues. As Steve Allen [1993] points out, they would have had less justification had the Bible been forthright in

159

condemning slavery, but, to the contrary, the Bible supports enslaving people other than your own. Indeed, Allen notes that not a single religion in the world condemned slavery until very recently; this shift in attitude is very possibly the result of secular, Enlightenment thinking, not religious thinking. If religions have changed their positions on slavery, they have done so in part because of the influence of secular philosophy.

Thus, it is not at all clear whether religion has been on balance a force for good or evil. Let us be charitable and say that, at least in the West, it has been a wash: that is, it has brought about no more evil than it has good. What is the origin of that evil?

The Book of Job

The Book of Job is often considered the classic study of the problem of evil. It is one of the three "wisdom" books of the Bible: the Book of Proverbs, the Book of Ecclesiastes, and the Book of Job. Here, wisdom means understanding the meaning of life, the purpose of existence, and the nature of good and evil: in other words, unanswerable questions for whose answers we constantly search. Bernhard Anderson, Professor of Old Testament Theology at Princeton Theological Seminary, [1966] relates the wisdom literature of the Bible to a long tradition of wisdom literature in the Near East and notes that the wisdom literature is very different from the prophetic portions in that it is concerned with individual existence rather than history or the giving of the Law. In 1966, he wrote that most of us are more at home with the wisdom literature than with the other portions of the Bible, not only because they are more universalist and we think of people as citizens of the world, but also because we have inherited our love of wisdom from the Greeks, Egyptians, and others. I am not so sure that he could write that today, but still I find these books, especially the Book of Job, very relevant to our topic.

160

Those who think the God of the Bible is always merciful should be given pause by the Book of Job. If we follow the Biblical literalists and take the book at face value, then a just God punishes a righteous man solely to win a bet.

No one knows when the Book of Job was written; Lane Fox favors some time between 400 and 300 B.C.E. The Book of Job consists of a short prose prologue and a short prose epilogue, which surround a long poem. Scholars assume that the prologue and the epilogue formed a well known legend and that a later author or poet wrote the poem that forms the bulk of the book. In this sense, the Book of Job is mostly a gloss on an old legend. Anderson and Nahum Glatzer [1969] assume further that the end of the poem was bowdlerized and that some other author or authors have rearranged the text and inserted several chapters (most particularly the speech of Elihu) to make the book turn out "right." We have no idea whether the author of the poem also transcribed the old legend or whether someone else added it to the scroll.

I am more interested in the being with whom God makes the bet. The Hebrew simply says "ha-satan," or "the adversary." I have found this word used a half-dozen or so times in the pre-exilic portions of the Hebrew Bible. Only in Zechariah 3 does it mean anything more than *an* adversary. (The King James Version and the Jewish Publication Society capitalize Satan in 1 Chronicles 21:1, but 2 Samuel 24 tells the same story with no indication of a supernatural being, so I take it to mean "adversary," not Satan, in Chronicles as well.) Thus, Job is nearly the only book in the entire Hebrew Bible where the root STN seems to mean a heavenly being (there is no indication that Satan is an evil being). Using "Satan" as a heavenly being is a Zoroastrian import, so we may accept that the original story of Job was not originally a Jewish story but rather a story that was circulating in the Middle East in post-exilic times.

The poetic insert that makes up the bulk of the Book of Job, however, is Hebrew poetry, and some of it is wonderful poetry. Unfortunately, it is also very difficult, and the JPS version has countless footnotes, "Meaning of Heb. uncertain," because so many of the words have been lost.

* * *

The original story of Job goes something like this: God tells Satan—brags, really—how pious is his servant Job. Satan's response is to remind God of Job's enormous wealth; take it away, Satan says, and he will curse you. God allows Satan to deprive Job, in stages, of his wealth, his family, and his health. Through it all, Job continues to worship God. Hence, presumably, the expression, the patience of Job (quoted in James 5:11).

Whoever coined that phrase did not study the Biblical Book of Job carefully enough. For, in the Biblical account, Job is visited by three "comforters." After seven days and nights (the period during which Jews sit *shiva*, or mourn for the dead), Job finally speaks. He is anything but patient but rather begins his speech by cursing the day of his birth. Throughout the rest of the book, the comforters take turns answering him, and Job replies to them in turn.

Eliphaz gives the first answer: For example, in 4:7, he asks, in effect, "Since when are innocent or upright men ever destroyed?" Later (4:8), he argues that those who sow evil reap evil. In short, Job must be guilty of something, or else he would not have been destroyed.

Job replies that he spoke recklessly and now asks God to show him his guilt, where he went wrong, how he has transgressed. Then he laments some more, allows that he is sick of life, and asks (7:17), "What is man that you [God] lavish so much attention on him?" In other words, Leave me alone!

The second comforter, Bildad, accuses Job of asking God to pervert justice. God does not despise the guiltless: in short, the same argument, but raised another notch in intensity. Job disagrees: God destroys both the guilty and the innocent. He implies (9:20) that God would twist any argument that Job might make and "prove" him guilty whether he was or not. He repeats (10:2) his request for an indictment. The third comforter, Zophar, accuses Job of babbling, argues (11:7 ff.) that Job cannot understand God, and advises Job (11:13) to repent.

Job's reply in Chapter 12 includes a paean to God's work, but that paean subtly shifts into a list of accusations against God: He exalts people and then strikes them down; he drives people crazy; he leads people astray. And, by the way, Job asks, how well would his comforters stand up to God's scrutiny themselves?

From here on, the arguments are repeated, but more vituperously and perhaps more personally, with Job alternately defending himself and lamenting. In Chapter 26, incidentally, Eliphaz comes very close to implying that all rich men are necessarily wicked: They engage in sharp business practices, deny sustenance to the poor, and send away widows empty-handed. He advises Job to throw himself to God's mercy. Job denies the charges and laments that he cannot get through to God to present his case.

Chapter 28 seems to be an added chapter on wisdom. In Chapter 29, Job asks, in effect, "Where are the snows of yesteryear?" and in Chapter 30 bemoans how he is ridiculed today, whereas he used to be honored.

Two of the comforters get to present their cases three times, but Zophar's third speech has apparently been replaced by the speech of Elihu, a young man who has been listening to the conversation. Anderson calls Elihu's speech an intrusion that was probably added by a later writer in order to state the case of orthodox (that is, standard or normative) Judaism even more strongly than had the comforters. In essence, Elihu says, "You comforters did not really answer Job. Here is the answer: God is greater than any man. How could God not be just?" Then he reiterates the argument that the wicked never prosper and adds that Job increases his sin by blaspheming. It is a pity that the words of the original poet, Zophar's last speech, seem to have been lost.

Finally, God loses *his* patience (Chapter 38). He appears to the assembly in a tempest (a whirlwind in the King James Version). In essence, God pulls rank on Job: Where were you during the Creation? Can you control the ocean? Do you make the sun rise? Job's answer, not surprisingly, is to acquiesce, and God lets him have it again (Chapter 40). Job admits that he spoke about things he did not understand, and he repents.

At this point, we return to the old legend. God berates the comforters, who did not speak the truth about him, as did Job; apparently, God has not yet read the Book of Job. At any rate, he restores Job's riches and his contentment, though no mention is made of Job's dead children nor his wife.

* * *

What are we to make of this story? First, if we take the book at face value, Job is *not* suffering for no reason. We know the reason, even if Job does not: because God has made a bet with Satan. That is, Job is suffering for a very bad reason. Biblical literalists explain away Job's suffering in the same way as Elihu: God's treatment of Job must necessarily be just, even if we cannot understand it, because God is necessarily just. This argument comes perilously close to saying that the ends justify the means or, more precisely, that some ends are so laudable that even the most reprehensible means are acceptable. That is, it was so important to prove Job's piety to Satan that God was justified in killing Job's family, destroying Job's livelihood, and torturing Job himself. It is hard for me to understand how ethical people can accept this argument when it is applied to God but not when it is applied to humans. Indeed, if you had been able to create an artificial intelligence, would you then be justified in killing it, just because you were its creator? If you answer no, then how is God justified in killing Job's family, just because he was their creator?

Glatzer [1969] notes that the Book of Job is well edited, and most readers will regard it as a single composition. If,

164

nevertheless, we omit the legend and concentrate on the poem alone, we still find no satisfactory answer to Job's questions. The Book of Job is more about people's responses to misfortune than it is about the existence or origin of evil.

Anderson says that the key to the story of Job is repentance: The book is fundamentally about Job's relationship with God, and the climax occurs when Job gives up his self-sufficiency and trusts in God. Others in Glatzer's anthology see Job as instantly apprehending the unity and goodness in the cosmos. In a moment of mystic ecstasy, Job recognizes the sovereignty of God. More bluntly, you could say that Job gives up trying to figure things out for himself and accepts God's judgement. But how does surrendering illuminate the problem of evil?

Jack Miles, [1996] a former Jesuit and now director of the Humanities Center at Claremont Graduate School, presents a circumstantial case that Job's last speech has been mistranslated and Job remains defiant to the end. Specifically, Miles thinks that the verb "abhor" in Job 42:5, 6,

> I have heard of thee by the hearing of the ear: but now mine eye seeth thee. Wherefore I abhor myself, and repent in dust and ashes

should not have been translated reflexively. Miles renders the same verses as

> Word of you had reached my ears, but now that my eyes have seen you, I shudder with sorrow for mortal clay

which he interprets as meaning, I mourn for the poor unfortunates who have to suffer under your dominion. To Miles, Job holds out and wins, and God therefore loses. God's attempt to shout Job down has failed, and God atones by returning Job's riches doubled. If Miles is correct, then the God of the Book of Job is indeed arbitrary and unjust, and his arbitrariness and

165

injustice are by no means excused or explained away by Job's concession and subsequent act of faith, as in the conventional view.

* * *

Kushner, [1981] by contrast, uses the Book of Job as a vehicle to illustrate his belief that misfortunes are simply misfortunes and are not directed by God. I cannot disagree with him but hasten to note that the misfortunes that befall Job are very explicitly ascribed to God, provided that we take the Book of Job as a single work (whether edited or not).

Nevertheless, I see the Book of Job rather as Kushner sees it. It is the story of a man's quest for understanding of reward and punishment, and perhaps of good and evil. Unhappily, the Book of Job does not truly illuminate these issues.

First of all, Job is a crybaby. He is rich and spoiled, unaccustomed to adversity. When misfortune happens to him, he wallows in a swamp of self-pity. A stronger man or a poorer man accustomed to setbacks would have just gotten on with his life.

Most important, though, the question of good and evil is never answered. When God finally appears, he belittles Job and trivializes his questions, but he does not answer them.

Now I will freely admit that, if God appeared to me in a tempest, I would be cowed into submission too. But, in the cold light of day, when God was gone, I would still have my questions and would still want credible answers. The scene where Job answers God—concedes defeat to God in the conventional view—reminds me of William Cummings's aphorism, "There are no atheists in the foxholes." That is, when your life is in immediate danger and there is nothing you can do to save yourself, you pray to God. The concepts of prayer and God are so deeply engraved in our psyches that I do not doubt that premise for a minute. If I were in a foxhole and bombs were dropping all around me, I would no doubt pray for deliverance, whether or not I really expected the prayer to work. You are not expected to behave or think rationally at all times. To conclude that, therefore, everyone deep down really believes in God is unwarranted, yet I have seen precisely that argument used by the clergy. Indeed, saying that there are no atheists in the foxholes (or that you will recognize the truth and come to believe in God on your death bed) is equivalent to arguing that you make the most sensible decisions when you are in no condition to make sensible decisions at all.

It is a dangerous argument, in any case, because there is a flip side to it. A woman I will call Ethel was the daughter of a Protestant minister who was the chaplain at a small, very religious college. When the minister died, Ethel's mother was inconsolable and apparently mourned for a longer time than Ethel thought was suitable. Ethel asked her mother why she was so distraught, since she expected to see her husband again soon. Her mother refused to answer, and Ethel concluded that her mother was not as firm in her belief as she claimed to be. I do not think that Ethel's conclusion was necessarily warranted, but the argument is every bit as sound as the foxhole argument.

What about Job? The next morning, in my view, he woke up and started asking the same questions. He kicked himself for not having had the presence of mind to ask God the questions he really wanted answered: Why is there suffering? Is it related to evil? I know that you created the world, but does that give you the right to torment people? I know that you make the sun come up in the morning, but what has that got to do with mercy? I don't beat my daughter; I protect her. Why, then, do you not protect me?

To paraphrase a famous remark of Alfonso X of Spain, "You created the universe, but if I had been around, I might have had some suggestions for you."

In short, unless you are willing to accept uncritically the argument that God knows best and you have no right to question his judgement, the Book of Job sheds no light on the questions of good and evil, reward and punishment.

167

The play, *J. B.*

The poet Archibald MacLeish wrote a modern version of the Book of Job: the verse play *J. B.* In the play, a couple of aging actors play the roles of Nickles and Mr. Zuss. (Old Nick is one of the names for the devil, and Zuss is obviously reminiscent of Zeus.) These actors watch as the story of Job plays itself out with all the determinism of a Greek tragedy, and they argue about Job's response. Nickles's predictions, like Satan's in the Book of Job, are always wrong, but, early in the play, he makes an interesting remark, "Not even the consciousness of crime to comfort him." Later, referring to their children, Job's wife tells him, "I will not / let you sacrifice their deaths / to make injustice justice and God good."

J. B. is concerned more with the original Job story—the prologue and the epilogue to the Book of Job—than with the comforters, but MacLeish's one scene with the comforters is a gem. I confess that I cannot distinguish among the comforters in the Book of Job, but in MacLeish's play they represent a Marxist, a psychoanalyst, and a priest, or their rough equivalents. Each tells Job that he is guilty, but for different reasons. The Marxist: You are guilty of belonging to the wrong class, living in the wrong century. The psychoanalyst: You are guilty of being guilty. The priest: Your heart and your will are inherently evil. Job keeps asking God to answer him, but he is without the self-pity of the Biblical Job. Finally, a Distant Voice calls to Job, using the language of Job 38-40, and Job repents. Nickles cannot believe that Job will accept his wife and his fortune back: Job knows the truth about God. He will fling his good fortune in God's face. But he does not.

Several times throughout the play, Nickles recites a little poem:

> I heard upon his dry dung heap
> That man cry out who cannot sleep:

168

"If God is God He is not good,
If God is good He is not God...
I would not sleep here if I could
Except for the little green leaves in the wood
And the wind on the water."

Here, when Nickles says "If God is God," I think he means, "If God is omnipotent," whereas "If God is good" means, "If God is benevolent." Mr. Zuss calls Nickles a bitter man, and he is, but the poem resonates well with the last lines of the play, when Job and his wife, Sarah, are combing through the ashes, and Sarah, finding a few leaves and petals, says, "You wanted justice, didn't you? / There isn't any. There's the world." Sarah tells Job that there is no justice, only love, and Job replies that God does not love; he is. "But we do," says Sarah. "That's the wonder." Justice and love do not come from God but take the place of God. In spite of the supernatural elements in *J. B.*, I see it as an essentially Godless play where, in the end, all that matters to us is our humanity. (See, however, MacLeish's essay on the Book of Job and Glatzer's [1969] introduction to that essay. In the essay, which predates the play, MacLeish concludes that God needs man's love, and only Job can prove that he loves God unequivocally. If MacLeish intended his play to be interpreted similarly, he has fooled not only me but also the critics.)

To Nickles, life is not worth living, except perhaps for the beauty in the world, as objectified by the leaves and the wind. For Sarah, life and love *are* the beauty in the world. To me, a much more satisfactory conclusion than that of the original!

The biological origin of evil

The Book of Job and the play *J. B.* do not come to grips with the problem of evil. The Book of Job avoids the issue, and *J. B.* is even less about evil than Job. What then is the origin of evil? To answer the question, let us look first at altruism. Where does

169

altruism come from? First, altruism is not limited to humans; it is found in other animals as well. The sight of a mother bird risking her life for her chicks is a well known example of altruism. Worker or soldier ants sacrificing themselves for the queen and the anthill are equally well known but less obvious examples. Why do these animals behave altruistically?

Richard Dawkins's [1976] selfish gene theory holds one possible explanation. In his book *The Selfish Gene,* Dawkins postulates that evolution takes place by the mechanism of survival of the fittest *gene,* rather than of the fittest *individual.* In other words, genes fight for survival, and the fittest genes are passed to the next generation. The animals that carry those genes are in some sense vehicles for propagating the genes. According to the novelist Samuel Butler, "A hen is only an egg's way of making another egg." Originally published in 1877, this epigram may contain more truth than its author knew. More precisely, however, Dawkins says that a hen is only a gene's way of making another gene.

Dawkins has been misunderstood, sometimes perhaps deliberately, because of his anthropomorphic description of a gene as selfish. [Stove, 1992] Others have called Dawkins and other sociobiologists racists because they do not like some of the possible implications of sociobiological theories. [Sperling, 1994; Wilson, 1995] For example, Susan Sperling, an evolutionary biologist with the University of California at San Francisco, reviews a book on the biological origin of morality. She describes its topic as "narrow biological determinism—a kind of genetic fundamentalism related to nineteenth century imperialist racism," but never shows the connection between sociobiology and racism, and admits that the author's politics are probably not much different from hers. Sperling has fallen into the trap of rejecting a biological theory of morality, not because of any evidence against it, but because she does not like it.

John Horgan, [1995] a staff writer for *Scientific American,* notes that men prefer young, healthy women who are able to bear children (this seems to be true across all cultures and is irrespective of the age of the men) and asks, "Do we really need Darwinian theory to tell us that?" I think that the answer is, emphatically, "Yes, we do," for otherwise we would not be able to understand why men who are undoubtedly not thinking specifically about procreation nevertheless generally prefer young women to old. One of the characters in Marge Piercy's novel *He, She and It* [1991] is a robot, and he is attracted to all women equally, irrespective of their ages. Piercy has had an interesting insight: The robot cannot procreate and is for that very reason indifferent to the age of a woman. Similarly, we might ask why young women sometimes are attracted to men substantially older than they, whereas young men are less often attracted to substantially older women. This question, too, may have an underlying sociobiological answer: because older men are still able to provide for the young women and their children, whereas older women are less able to bear children.

In addition, Dawkins's calling an animal an automaton that blindly carries out the commands of its genes seems a bit overstated. Dawkins's argument, however, is fairly temperate and boils down to this: The term "selfish" gene is a metaphor that means not that each gene fights for its own survival, but rather that all the (identical) replicas of a single gene behave *as if* they are fighting to pass as many as possible of those replicas to the next generation. They do so within the context of a myriad of other genes that are doing the same thing. And they do so because only those genes that succeeded in the past have survived to the present. In other words, Dawkins's theory is much like Darwin's theory, but it is applied to sets of genes, rather than to individual organisms. Darwin did not claim that organisms consciously fight for survival, and Dawkins does not claim that genes do so. Dawkins's theory underlies Darwin's, however, in that it attempts to explain the *mechanism* of evolution, much as the atomic theory underlies the gas laws and explains the mechanism of gas behavior.

I do not know whether the selfish gene theory has universal validity, but it certainly has the potential to explain what would

171

otherwise be inexplicable behavior. If evolution were based on survival of the fittest *organism*, then the mother bird's instinct to protect herself would be stronger than her instinct to protect her young. Why then is the reverse true? Dawkins would say it is because the mechanism of evolution is survival of the fittest *gene*, and the young represent her genes' investment in their future. The mother bird that is driven to protect her genes more than herself is therefore more likely to pass her genes to succeeding generations than the mother bird that abandons her young to save herself. More convincingly, why should not all female ants struggle to become queen? Because the females in the hill share most of their genes with the queen. It therefore makes little difference which ant is the queen and which die defending the anthill. The genes survive.

Much of what we call altruism in humans can probably be explained in much the same manner. For example, we are more likely to sacrifice for our children than for our nephews. More likely to sacrifice for our nephews than for our compatriots. More likely to sacrifice for our compatriots than for foreigners. Altruism toward relatives is often ascribed to *kin selection*, and altruism toward members of the same group is ascribed to *group selection*. Group selection is controversial, and Dawkins flatly denies that it exists. I will mention it briefly below. At any rate, kin selection means that natural selection favors those organisms that are altruistic toward their relatives; your genes have a better chance of survival if you are more altruistic toward your relatives than toward complete strangers, because your relatives have more replicas of your genes than do complete strangers. That is not to say that all altruism has a biological explanation nor that all human behavior is biologically determined. Rather, I am suggesting that altruism is not limited to humans and, in humans, may well have a strong biological component.

Other animals, such as chimpanzees, groom each other or otherwise come to the aid of nonrelatives without immediate reward. This is sometimes called *reciprocal altruism* and is distinguished from cooperation in part because reciprocal

172

altruism, in contrast to cooperation, has a delayed payoff. [de Waal, 1996] Theorists have begun to investigate the origin of reciprocal altruism and cooperation, and find that these sometimes replace or supplement competition.

Let us begin with a game called the Prisoner's Dilemma. [Waldrop, 1992; Dawkins, 1989; Dennett, 1995] Two men have been charged with a crime and are held in jail in separate cells. Each is given the same offer: If you confess and implicate your partner ("defect"), we will not only free you but also offer you a substantial reward; your partner will go to jail and, in addition, will be fined heavily to cover the cost of your reward. The offer stands, however, if only one of the prisoners defects. If both prisoners remain silent ("cooperate"—that is, with each other, not the police), the police will have no evidence, and both prisoners will go free. If both defect, they will both go to jail, but neither will have to pay the other's reward. Neither prisoner can trust the other to cooperate, and each realizes that, if one defects, the other will pay more dearly than if both defect. In the end, therefore, both defect and both go to jail, in an effort to minimize their losses.

Mitchell Waldrop notes that real-world decisions are rarely as stark as the choices available to the prisoners but considers the Prisoner's Dilemma depressingly accurate about the need to protect yourself against betrayal. An animal that is too trusting gets eaten; a person who is too trusting or too altruistic gets robbed or taken for a sucker. Why, then, do unrelated animals ever cooperate?

Part of the answer is found in repetitive iterations of the Prisoner's Dilemma. Suppose you get to play the Prisoner's Dilemma over and over with the same opponent. After a while, you will acquire some data and learn whether to trust that opponent. Iterated Prisoner's Dilemma is therefore more like real interactions among organisms in bands or societies.

The political scientist Robert Axelrod in the 1970's organized a now famous tournament in which each participant played 200 consecutive rounds of Prisoner's Dilemma with each other participant. The participants were not people but rather different strategies that had been submitted to Axelrod, who wrote a computer program that allowed each of 15 strategies to compete with each other strategy. Entrants were awarded points depending whether they cooperated or defected. For example, if you defect but I cooperate, I get no points; you get 5.

The winning strategy was the simplest: Tit for Tat, submitted by the game theorist Anatol Rapoport of the University of Toronto. Tit for Tat begins by cooperating; thereafter, it duplicates the opponent's previous move. It is cooperative but tough. It never defects first. If you cooperate with it, it will cooperate with you, but if you defect, it will punish you by defecting on the next round.

The implication for biology is clear: Suppose that a few tough cooperators arise by chance somewhere. They use a strategy like Tit for Tat in their encounters with other individuals. Once they meet, they will cooperate with each other and outperform their more selfish rivals. If they pass their trait on to their descendants, then these tough cooperators will eventually prevail. (Remember, in our discussion of the eye, we found that even minuscule changes can accumulate over many generations and result in something very different from the starting point.) If they are also endowed with memory, then the cooperators will begin to recognize their friends (defined as those who cooperate with them) and not only cooperate, but finally engage in reciprocal altruism, that is, altruism with no immediate gratification.

* * *

Frans de Waal, [1996] a researcher at the Yerkes Regional Primate Research Center in Atlanta, finds the rudiments of morality, not only in primates but also in cetaceans; both of these groups have been documented to act altruistically. De Waal thinks that monkeys and apes, as well as possibly other animals such as elephants and whales, are keenly aware of the needs of their conspecifics and are concerned for them. He does not discuss the mechanism whereby the apes acquired this concern but notes that human morality or ethics may well have evolved from similar attributes. He suspects that evolution provided us with the prerequisites for morality and that societies, including animal societies, provided the need for morality. He does not argue that morality is built in, but rather sees moral decisions as societal decisions, and notes that some societies have moralities that are diametrically opposed to those of others, as we in the United States have subcultures that favor abortion and others that oppose it and regard it as murder. Morality or ethics, in other words, may not be strictly biological but may be taught by cultural, societal, or other imperatives. De Waal sees the mind not as a *tabula rasa* but rather as a "checklist with spaces allotted to particular types of incoming information." He speculates that we are born with a moral ability that is analogous to our language ability. As we grow up, we figure out our native moral system just as we figure out our native language. Indeed, the biologist Lyall Watson [1995] suggests that, just as there is a critical time for learning language, there may be a critical time for learning morality: Miss it, and you are lost.

De Waal argues that cooperation leads to reciprocal altruism. You have to have a good memory to engage in reciprocal altruism, since you have to remember who helped you, and you also have to speculate who will help you in the future. Hence, we see reciprocal altruism primarily among more-intelligent species. If you also have sympathy for others (empathy is not enough; you have to be affected by what affects another), then you may develop true altruism, that is, altruism with no wish for reward. Thus, de Waal says that cooperation + memory → reciprocal altruism, but reciprocal altruism + sympathy → (true) altruism. The philosopher Moses Maimonides called giving anonymously to someone you do not know the highest form of charity, and that indeed seems several steps removed from reciprocal altruism.

175

De Waal also notes, depressingly, that morality may be a luxury that only the affluent or well-fed can afford. In primate societies, the rate of starvation and malnutrition among low-ranking offspring is high. Altruism among primates is mostly if not entirely reciprocal altruism, not altruism for its own sake. In addition, de Waal cites a case study of an African tribe, the Ik, by Colin Turnbull. Turnbull never saw anyone over the age of 3 being fed by anyone else, and he noted people eating away from their homes to avoid the need to share. The Ik evidently felt no sympathy for others and some stole food literally out of the mouths of weak or elderly people. Turnbull saw them laugh heartily when a child burned herself, for example.

De Waal decries the reductionist approach of Dawkins, as well as the claims of others that all apparently altruistic actions are at root selfish, though he agrees that altruism can be rewarding in the sense that it makes you feel worthy. He notes that you can easily thwart the "desires" of your genes, and he is surely right; people who decide not to have children destroy any "hope" of their genes for survival. Nevertheless, for our purpose, de Waal comes to a similar conclusion to that of Dawkins when he argues that the capacity for morality is inherent in our biological makeup.

Thus, we see that degrees of altruism exist among many animals, and altruism has a plausible biological explanation. Explaining altruism or morality requires no special hypothesis such as the existence of a supernatural being.

* * *

Now, what about evil? Kushner uses evil to include misfortune, and I think that he deals with misfortune very well. But the real problem for believers is true evil: that which is morally reprehensible and causes deliberate harm. Why is there evil?

Once, on a science program on television, I saw some harlequin shrimps carrying off a starfish. (There were at least

176

two, and they were cooperating, by the way.) According to the narrator, they were going to take the starfish to their lair and eat it over the next several days, starting from the ends of the legs and working inward. The starfish was alive and would remain so for most of the ordeal. I think that most of us would perceive that as kind of "gross," but hardly anyone would consider the shrimp to be evil. Rather, we would say, they were engaging in biologically determined behavior, much like the ants who sacrifice themselves for the good of the colony. Similarly, when a mother guppy eats her young, when a cowbird lays her eggs in a warbler's nest, and when a male lion kills the offspring of another male, no one calls any of them evil.

* * *

Do humans engage in biologically determined behavior? It certainly seems so, though there are so many other factors that it is hard to argue that we are programmed like robots or insects. Nevertheless, studies of identical twins, especially identical twins reared apart, have shown striking similarities in behavior. [Holden, 1980; Wright, 1995] For example, Lawrence Wright cites a study of the rate of criminality among adopted children. According to the study, the chance that a child will be a criminal increases from about 3 percent when neither the adoptive nor the biological parents are criminals, to 7 percent if the adoptive parents are criminals, to 12 percent if the biological parents, but not the adoptive parents, are criminals. Very roughly, the effect of biology on causing criminality is twice that of environment. That is not to say that there is a gene for criminality. Rather, as Winifred Gallagher [1994] notes in *The Atlantic Monthly*, a person with a certain personality type in one environment might become a criminal whereas, in another environment, the same person might become a famous adventurer or a test pilot. The adventurous behavior, nevertheless, seems to be largely biological in origin, even if the manner in which it is expressed is culturally determined.

177

Wright cites other evidence that adopted children eventually grow up to resemble their biological parents more than their adoptive parents in both IQ and personality. Gallagher presents evidence, based on studies of monkeys, that inhibited infants raised by an uninhibited mother can become uninhibited also (although it does not seem possible to inhibit an uninhibited infant simply by raising him or her with an inhibited mother). The mechanism for the change of temperament is chemical and is related to the level of a specific neurotransmitter in the infant's brain. Wright argues similarly that the individual is not programmed by his or her genes but rather that the genes determine the way in which the individual responds to the environment. Thus, says Wright, as they grow older, identical twins reared apart grow more like each other and more different from their adoptive families because of the way the environment influences them. This seems to me a distinction without a difference: Either way, biology plays a major role in behavior, and it seems as if your underlying temperament is determined by biology. Even the monkey experiments cited by Gallagher establish that the change of temperament in the infant has a biological origin.

In 1975, the anthropologist Joseph Shepher reported on a study of children in the Israeli kibbutzim. [Tiger and Shepher, 1975, p. 7] In certain of the kibbutzim, the children were reared communally in a "children's house." Shepher recorded 3000 marriages of kibbutzniks over three generations. He found not a single marriage between adults who had been reared together between the ages of 3 and 6. Tiger and Shepher did not have a ready explanation for this fact but noted that it violates the *law of propinquity*, a sociological principle that people marry others who are geographically close and have shared similar experiences, that is, that boys often marry the girl next door.

I think we can now suggest an explanation. A great many animal species have mechanisms for avoiding incest. [Thornhill, 1993] In some species, one sex or the other leaves home and finds a mate elsewhere. In other species, however, close relatives simply refuse to mate. Some observers believe that relatives avoid mating with other relatives not specifically because they are relatives but because they are familiar. That is, these animals do not know who their relatives are. Rather, they avoid incest by refusing to mate with animals whom they identify as familiar by their size, appearance, or odor. They may even refuse to mate with cousins if the cousins, for example, smell a bit like their siblings or give off similar pheromones. This may be adaptive behavior that reduces the likelihood of genetic defects due to inbreeding.

In many species, animals that are reared together are necessarily relatives. This is so among rodents that have large litters but do not live communally, for example. Shepher's study suggests that the human prohibition against incest very possibly has a biological origin, because adult humans who had been reared together as children rigorously avoided mating with one another, even though they were not related. In short, incest may not be inherently evil but may rather be a biological adaptation that has expressed itself in cultural taboos.

Why then do we consider it evil when one human engages in incest, kills another, cheats another, lies to another? No less than other animals, those humans may well be engaging in biologically determined behavior: Survival of the fittest sometimes means cooperation, but it sometimes means no holds barred. Why, then, are humans who bar no holds evil? I think you can argue this way: because humans, almost uniquely among the animals, are able to put themselves in the place of another. Because we can communicate effectively and therefore can recognize the pain of another. We can sympathize with others, that is, feel what others feel and let it affect us as well. Most of those things that we consider evil involve inflicting pain on another or bringing misfortune to another. Why are these things evil in humans but not in other animals? Because we say so! That is,

Humans decide what is evil and what is not.

179

You could similarly try to account for the existence of moral codes sociologically or anthropologically. Such an explanation is not incompatible with a genetic explanation and may well be related to group or kin selection. Group selection has long been in disrepute but is making a comeback. [Lewin, 1995] It seems "obvious" to some that groups must necessarily select members who will cooperate, or else the group will lose cohesiveness and eventually collapse in civil war. Presumably, members who are not cooperative or reciprocally altruistic are either ostracized or killed. Since those members do not propagate their genes, the group becomes more altruistic with time. The evidence to support this hypothesis, however, has been lacking, and mathematical models suggest that it is not true. For example, group selection predicts a population that supports more females than males, since the number of females ultimately determines the number of offspring and therefore the success of the group. In most sexually reproducing animal species, however, the ratio of males to females is about even.

* * *

Recently, the evolutionary biologist David Sloan Wilson of the State University of New York at Binghamton has revived the idea of group selection. Wilson argues that group selection can take place when groups or societies compete, though he grants that it cannot work on societies in isolation. When groups compete, those whose members cooperate survive the competition best. (The survival of the Jews, which I discussed above (see "Signs," Chapter 3), may be no more than a striking example of group selection. [Young, 2000] It is curious, therefore, that the Jewish civilization was nearly destroyed in the years 67-70 C.E., when the Jews did not fight the Roman siege of Jerusalem effectively because they were simultaneously involved in an internal civil war.)

According to this view, the most internally cooperative societies flourish. No complex society can have its citizens constantly at each other's throats, either literally or figuratively. If you can get people to internalize a moral code, you can effect a degree of social control more effectively than you can with an imposed legal code. Thus, a moral code may well have evolved because of the exigencies of complex societies.

In early societies, the moral code extended only or primarily to the members of the tribe. Thus, societies that forbade stealing, lying, murder, adultery, and so on applied their strictures mostly to themselves. In the Bible, we see many examples where the Hebrews or other tribes wipe out this enemy or enslave that enemy. Indeed, in 1 Samuel, Saul is chastised for not obliterating the tribe of Amalek. Our own society suspends the commandment not to commit murder during war by redefining murder to exclude killing an enemy soldier or, in certain circumstances, a civilian. We suspend the prohibition against slavery by importing merchandise made by foreign workers who are little better than slaves. We suspend the commandment to love our neighbors when we seal our borders against the poor of other nations. The fact that the moral code is not universal but extends primarily to the members of our own group adds force to the argument that morality is related to group selection. If it were not, we might apply our moral code as rigorously to others as to ourselves.

I will not go farther with this argument, in part because I cannot authoritatively choose between the proponents and the detractors of group selection. Some form of group selection, or survival of the fittest tribe, nevertheless offers at least a plausible explanation of the existence of moral codes. Like the explanation based on genetics, such an explanation requires no supernaturalism. Indeed, a biological explanation and an anthropological explanation may complement each other. We are, after all, biological creatures who live in societies. My point is not to explain in detail where we get our moral code but rather to note that it is unnecessary to invoke the supernatural to find

181

the origin of morality. It is probably within us or our socialization. We decide what is moral. We decide what is evil. That we do not always agree shows only how difficult it is to make moral decisions when there is a welter of conflicting factors. Evil, like beauty, is in the eye of the beholder.

* * *

All societies are said to have a variation of the Golden Rule (I will ignore for the moment that they apply it more frequently to themselves than to outsiders, though the fact is not insignificant). In the Hebrew Bible, the rule is stated, "Love your neighbor as yourself" (Leviticus 19:18). Rabbi Hillel (1st century B.C.E.-1st century C.E.) stated it in the negative, "Do not do to others what is distasteful to you." Matthew (7:12) and Luke (6:31) state it, roughly, "Do to others as you would have them do to you." Did this rule come from God or from humans? We can make plausible arguments with supporting evidence from biology or anthropology that it came from humans: perhaps because we feel compassion for one another, or because societies needed it as a means of forcing people to coexist peacefully, or both. That it came from a higher authority is an unnecessary assumption without supporting argument.

The existence of evil is not a problem for scientists; it is a major problem for theists, if not the major problem. Altruism is likewise not a problem for scientists. Both evil and altruism can be explained by evolutionary biology. No matter its origin, however, evil seems to me to be a fatal flaw to literalists who claim that God is benevolent. The existence of evil does not disprove the existence of God, but it surely calls into question his benevolence and, I think, wholly discredits the idea of a benevolent God.

Conclusion

Let us stop for a moment and take stock. We have seen that unsupported beliefs are little more than hunches; only beliefs supported by empirical evidence should be accepted as true. Further, beliefs that stand in opposition to well founded scientific facts may be regarded as simply wrong, no matter how venerable they are. The evidence provided by miracles, signs, and literal readings of the Bible cannot stand even the most casual scrutiny. The existence of evil is a serious argument against a benevolent God, if not God in general. In stark contrast, the existence of altruism does not count much in favor of a benevolent God, since we can explain altruism without recourse to God. Thus, the evidence is weighing in against a benevolent God, if not against God himself. What do philosophers and theologians have to say? That is the subject of the next chapter.

Why does the universe go to all the bother of existing?

STEPHEN W. HAWKING

Chapter 6
Aquinas's error

I will use the Roman philosopher Lucretius (ca. 100-55? B.C.E.) as an archetype of a certain kind of philosopher and, I hope, not as a straw man. As I reread his work *The Nature of the Universe*, I remembered why I had been so impressed with it when I first read it. Lucretius was a keen observer, and many of his arguments are brilliant. But many if not most of his conclusions about the physical universe were dead wrong, and his belief that the universe was composed of atoms was wholly unsupported by evidence. Yet Lucretius thought he was describing the universe accurately and in detail. I want to examine Lucretius carefully because I see many of the same flaws in much later philosophic and religious writing: The deductions either are not testable or do not follow from the evidence.

Lucretius [1951] inherited the atomistic theory of Democritus and Leucippus through his mentor Epicurus. Lucretius was a materialist and believed that all knowledge could be obtained by using his senses, though he recognized that there were some things our senses could not detect. He believed that everything in the universe was made of atoms and empty space. That space was empty was "obvious" to Lucretius; if it were not, then the atoms would be immovable.

Atom comes from the Greek roots *a*, not or without, and *tom*, cut, as in microtome or tomography. That is, atoms were uncuttable or indivisible. Some of Lucretius's atoms, however, were more like our molecules, and some corresponded to our cells. They were indivisible only in the sense that, if you cut them, they are no longer the same atoms; this is true of what we

185

today call atoms as well. Lucretius believed that there were atoms of stone, atoms of air, atoms of light. His atoms could not be created out of nothing, nor could they be destroyed. Atoms could join but not in any conceivable way, because, otherwise, there would be all sorts of odd hybrids or chimeras. Rather, everything grows true to form. When we eat, atoms are dispersed throughout our bodies according to certain laws. Similarly, all matter moves according to laws. The atoms of pepper are sharp and thorny; the atoms of water are spherical to allow the water to flow easily; and the atoms of stone are hard and interlocking.

Lucretius made very astute observations. For example, he deduced that the atoms themselves must be colorless. This is so because white materials, such as foam, can be created from materials that are not themselves white. Further, the objects are not visible unless they are in the light, and certain objects can appear different in different lighting. On the other hand, he thought that we see because atoms are perpetually emitted from the outer skin of an object and into our eyes. His explanation of why light is necessary is especially convoluted. Neither Lucretius nor any Greek philosopher understood clearly how light was necessary for vision.

Lucretius argued that sentient animals must be composed of insentient matter. He thought that mind and spirit must be composed of matter, though the mind must be composed of exceptionally small atoms, because it operates very fast. Indeed, the atoms are spherical, like water atoms, or else they would not be as mobile as they are. The atoms of the spirit are also very small and light, because a dead body loses no measureable weight when its spirit departs.

Lucretius deduced that the universe was not run by the gods. He argued that the universe was infinite; in his mind, no single entity could possibly rule the measureless or be in all places at all times. He believed in the birth and death of the soul and thought that there was nothing after death. He subscribed to the Epicurean philosophy, live life to the fullest, for there is no other. That philosophy is sometimes incorrectly assumed to be a philosophy of hedonism, but to Lucretius it meant study and the intellectual life as well as what are more commonly called pleasures. Much of the Biblical Book of Ecclesiastes seems based on Epicureanism, though the author ultimately rejects Epicureanism as he rejects every other worldview he examines. I find much to be admired in the Epicurean worldview.

Like the Greeks before him and like many philosophers after him, Lucretius thought that anything he deduced was correct. Many times he assures us that what he tells us is the truth and that, once he explains it, we will understand. Nevertheless, much of what Lucretius believed he had deduced is plainly wrong. His belief that we see because atoms are emitted from objects and are intercepted by our eyes, for example, is no longer defensible. We know today that we see because objects scatter or reflect light and the light gets into our eyes. In addition, if Lucretius were correct, then all objects would gradually lose weight, and they do not.

We call Lucretius, Democritus, and Leucippus philosophers, though they were also, in some sense, natural scientists. The word scientist was not coined until 1834, but the distinction between natural scientist and philosopher began to take shape, say, at the time of Galileo. Galileo was one of the first people to perform a scientific experiment, as opposed to a careful observation. His experiments, rolling and sliding masses down inclined planes, conclusively disproved Aristotle's claim that heavy bodies fall faster than light bodies. From that time on, science proceeded very differently, and scientists rejected or were very skeptical of deductions that were unsupported by evidence. Only in that way did they progress beyond the unsupported deductions of Lucretius and other philosophers to mature sciences. However logical they appear, Lucretius's deductions about the nature of the universe cannot stand up to scientific scrutiny. In the following sections, I will suggest that certain philosophical deductions about the existence of God cannot stand similar scrutiny.

187

Elemental God

I have been very careful until now not to define what I mean by God. That was not an oversight. Rather, I have tacitly adopted the traditional Western view of God as a benevolent, omnipotent, and omnipresent, as well as "infinite," being who may or may not have somehow limited himself to give us free will. I have done so because that is the implicit view of most of those whose works I have examined and, I assume, of most laypersons today, whether they are believers or not.

Now, however, I am about to infringe on the turf of professional philosophers, so I had better be especially careful. Thus, let me define God this way:

> *God is the purposeful and intelligent creator of the universe.*

I almost said conscious, but I do not really know what that means. Indeed, I am not entirely sure I know what I mean by intelligent, and I may be anthropomorphizing when I ascribe "purpose" to God as well. The main point is that I want to define God as narrowly as possible but include the idea that, if God created the universe, he did not create it reasonlessly. My definition can accommodate the beliefs of deists, for example, who believe that God set the world in motion but left it to run itself, and theists, who believe that God is continually active in the operation of the universe. I do not want to complicate the arguments with extraneous considerations like whether God is just, looks over us, or occasionally intervenes and performs miracles. Nor am I concerned right now with whether God is supernatural, outside the universe, or infinite, or with whether he can violate natural law when it pleases him to do so. Better to decide about God's existence before we worry about his properties, activities, or habitat.

What does philosophy have to say to a scientist—indeed, an experimentalist—like me? For an answer, I began with the

188

college textbook *Philosophy of Religion: An Anthology*, edited by Louis P. Pojman. [1987] Pojman is Professor of Philosophy at the University of Mississippi and was the moderator of the discussion between Moreland and Nielsen. [1993] His book is a series of readings by influential philosophers, joined by introductions by Pojman. The readings are arranged specifically so that they rebut one another and give a relatively unbiased overview. I will examine each argument briefly from the point of view of an empiricist and bring in scientific facts or theories wherever possible.

The Ontological Argument

The argument was developed by St. Anselm (1033-1109 C.E.), but the word was not coined until 1877. It comes from the Greek roots *ont*, being, + *logy*, doctrine or theory. Thus, the Ontological Argument is a theory of existence. It is an *a priori* argument, that is, a purely deductive argument based on reasoning or supposedly self-evident propositions, not on observed facts.

The Ontological Argument goes something like this: [Pojman, 1987; Hick, 1964]

> *Imagine a being that is so great that there can be none greater (the greatest possible being, for short).*
> *If the being did not actually exist, then it would not be the greatest possible being.*
> *Therefore it must exist in reality, not just in your imagination.*

The argument assumes that God is a "being" without precisely defining being and skirts dangerously close to anthropomorphism. How do we know that God is a being in the usual sense of the word? In addition, the second step depends on

189

the unstated proposition that existence in reality is "greater" than existence in imagination.

There are many other things wrong with the Ontological Argument. Greater is not defined, though we can take it to mean more nearly perfect. It is not obvious why or how existence in reality is greater than existence in the imagination. Maybe I *can* imagine a being that is greater than a being that actually exists; I can easily imagine a whale that is larger than any whale that actually exists. In addition, there is no reason that the greatest existing being has to be the creator.

Suppose that I said, "Imagine an emerald that is so green that there can be none greener," and continued the argument exactly as above. An emerald that really exists must be greener than an emerald that is only imaginary. If it did not exist, then it would not be the greenest possible emerald. Therefore it must exist in reality and not just in your imagination. You would argue that there is a limit to how green an emerald can be, and in fact a great many emeralds might have precisely the same greenness. There is no such thing as infinite greenness. Yet the Ontological Argument proceeds in the same manner and leads you to believe that there must be at least one infinitely great being, simply because you can imagine such a being. Monotheists assume that there is exactly one such being, though the Ontological Argument does not imply that.

Anselm refuted this line of thought by arguing that the logic pertained only to the greatest possible being. This being, by definition, must have existed for all time and must continue to exist for all time. You can easily conceive of any material object as not existing. But can you conceive of the greatest possible being as not existing, even in your imagination? That is a little like saying, "Try *not* to think about a purple elephant with pink polka dots." The minute I say it, a purple elephant with pink polka dots pops into your mind. You will try to banish it without success, partly because the act of banishing requires you to think about what you are banishing. Similarly, as soon as Anselm says "greatest possible being," such a concept pops into your mind.

If, however, you can hold the concept of a greatest possible being but can simultaneously conceive it as not existing in reality, then Anselm's refutation fails. The Ontological Argument is revealed as circular: If you can conceive of a greatest possible being, then it must exist. But if you can conceive it as imaginary, then it need not exist.

Rather than talk of material objects, let us consider another kind of abstract entity. Suppose that I said, "Imagine a number that is so large that there can be none larger," and similarly continued the argument. You would argue that there is in fact no such number; numbers can get large without limit. Further, I argue that infinity is a concept that applies only to numbers and other abstract constructions. The number of points on a line is infinite, but the point itself is an abstract construction that has no physical reality. No elementary particle (electron, proton), for example, is truly a point. Similarly, no real, physical quantity can be truly infinite. Indeed, scientists have expended a great deal of time and effort trying to rid their theories of apparent infinities. Infinity is a mathematical construction that does not apply to the real universe. As far as we know, even the universe is finite. The number of electrons or quarks in the universe, though very large, appears to be finite. No matter how old the universe becomes, it will never be infinitely old. Thus, an argument that involves infinity can have no physical reality but is at best an approximation to reality.

Another major problem with the Ontological Argument is that it is not testable. It is based on no evidence, and it concludes with no testable result. The same is true of many, though not all mathematical theorems; you can experimentally test, say, the theorem of Pythagoras and show that it is true within experimental uncertainty. Indeed, your ability to do so verifies that space is Euclidean (flat); if space were not Euclidean, then Pythagoras's theorem would not be true in the sense that it would not correspond to physical reality, even though the theorem is absolutely correct in the abstract. That is, Pythagoras's theorem depends on the assumption that space is

191

Euclidean. Given that assumption, we can show that the theorem holds in the real world. If, however, space were not Euclidean, then the theorem would not be objectively true in the real world, however abstractly true it is an abstract Euclidean space. Indeed, if we were to draw a very large triangle on the surface of the earth, we would find that the theorem did not hold, because the earth's surface is not flat. Thus, the fact that the Ontological Argument is a deductive argument does not excuse it from the requirement that it be testable if it claims to describe reality.

Other theorems, like those of algebra, may be useful even though they cannot be tested experimentally. No one, however, claims any kind of physical reality for the results of those theorems, the way Anselm claims physical reality for the outcome of the Ontological Argument. The Ontological Argument seems unique in that it is an abstract argument, but its supporters claim that it *necessarily* describes reality even though they cannot demonstrate the truth of their claim. Like Lucretius, Anselm evidently thought that whatever he deduced about the external world was correct. Unlike Lucretius, Anselm had an ax to grind: He firmly believed in God and set out to prove God's existence, not to examine the arguments critically. In a literary allusion to the Psalms, he called nonbelievers fools and intended to convince them with what he thought was an impervious argument (Psalms 14:1 and 53:1). The Ontological Argument is a polemic, not a reasoned argument.

* * *

Paul Davies is Professor of Mathematical Physics at the University of Adelaide in Australia. In 1995, he was awarded the Templeton Prize for Progress in Religion. In his book *The Mind of God*, [1992] Davies expresses sympathy with the Ontological Argument, while conceding that it "reeks of logical trickery." Following Kant, [Pojman, 1987; Hick, 1964] Davies argues that one problem with the Ontological Argument is that it treats existence as a property, like the color of an emerald or the

192

magnitude of a number. Davies argues that existence is not a property, like color or mass; you have to know that something exists before you can assign properties to it. As Davies puts it, he can speak about having five small coins and six big coins, but can he have five existent coins and six imaginary or nonexistent coins?

Is existence a property? I think not, but I well remember my philosophy professor, Colin M. Turbayne at the University of Rochester, telling an anecdote about how he asked his mother-in-law to visualize an apple. Now, he said, imagine the apple without its color. OK. Imagine it without its smell or taste. OK. Without its substance or texture. Harder, but OK. What have you got left? The apple! Evidently, to Turbayne's mother-in-law, the apple existed independently of any of its properties. Existence was the only property it had.

Turbayne's apple was not a real apple. It existed only in the mind of his mother-in-law. But I think it would be more accurate to say that what really existed in her mind was the linguistic construct called an apple. An apple with no properties does not exist in the sense that it has no objective reality.

Anselm has the cart before the horse. A being that does not exist is not imperfect—it is nothing. You can, for example, make the statement that, if triangles exist (and space is Euclidean), then the sum of the angles of a triangle is 180 degrees. That statement is demonstrably correct. But you cannot use it to prove that triangles must necessarily exist; there is no reason that triangles have to exist. In fact, triangles do not exist, except as abstract or imaginary entities. A triangle whose angles add precisely to 180 degrees is an abstraction made of perfect line segments that lie in a perfect plane. The only "triangles" of which we have experience are approximations that are drawn with imperfect line segments on imperfect planes, and whose angles add only approximately to 180 degrees. Yet I can easily conceive of a perfect triangle, while recognizing clearly that it does not exist.

Similarly, you can claim that, if God exists, then he is a perfect being. Knowledge of this property, however, does not guarantee God's existence, since the claim is only hypothetical. There is no reason that God has to exist, and the conclusion that he is perfect has substantially less rigor than the theorem about the triangle.

As Davies puts it, it is not obvious that logical propositions, such as the Ontological Argument, necessarily relate to the real world. That is, if Anselm's greatest possible being exists in the realm of pure logic, must he necessarily exist in the real world? If you are tempted to answer "yes," then Davies invites you to consider "the success of mathematics in describing nature, especially the underlying laws." This success has suggested to some that mathematics—which is nothing more than a collection of logical propositions—underlies the entire universe.

Davies calls this line of thinking "badly misconceived," and I cannot disagree with him. If you can necessarily derive a physical law from a mathematical or logical proposition, then virtually any mathematical theorem ought to imply a physical reality. That proposition is simply not true. The Law of Gravity, for example, is an inverse-square law. That is, the gravitational force between two distant bodies is inversely proportional to the square of the distance between them. In mathematical language, $F \propto 1/r^2$, where F means force and r is the distance between the two bodies. As far as we can tell, the exponent is exactly 2; no other value, not even 2.000 001 or 1.999 999, gives results consistent with astronomical observations.

You could, if you wanted to, develop a theory based on an inverse-cube law or any other law you chose. It could be logically and mathematically consistent. When you tested the law against astronomical observations, you would find that it did not describe physical reality. The Ontological Argument leads to no testable conclusions and has therefore never been tested. We have no way of knowing whether it is correct. Merely stating it and showing that it is logical is no proof that it corresponds to reality. Indeed, those who take the Ontological Argument at face value are falling into the same trap as Lucretius: believing that what they deduce is necessarily correct.

* * *

My own view of the Ontological Argument is that it is an extrapolation from what is known to what is unknown, and such extrapolations are extremely dangerous. By way of example, consider a light meter that develops a voltage in response to being illuminated with light. At low intensity, the voltage is proportional to the intensity of the light; that is, if we double the intensity of the light, then the voltage also doubles. If you worked only at low intensity, you might conclude that the voltage was always proportional to the intensity, no matter what the intensity. Call this the "law of proportionality."

In reality, however, there is a limit to the voltage that the light meter can develop. That is, the light meter ultimately *saturates* in that doubling the intensity yields less than double the voltage. At first, the voltage is only slightly less than doubled, and the law of proportionality remains approximately correct; as the intensity is increased further, however, the voltage deviates more and more from the prediction of the law of proportionality. Eventually, the voltage approaches an upper limit, and an increase in intensity yields only an immeasurable change in voltage. The law of proportionality, however, would have led you to believe that the voltage would double as long as the intensity was doubled, no matter how high the intensity. That is, the law of proportionality says, incorrectly, that the voltage will increase without limit if the intensity increases without limit. In scientific shorthand, we say that (according to the law of proportionality) the voltage will be infinite if the intensity is infinite, even though we know that we cannot really achieve infinite intensity.

In place of the law of proportionality, we could develop a more complex mathematical expression that describes the relationship between intensity and voltage; call it the law of

195

saturation. The law of saturation is closer to the truth than the law of proportionality, and it says that you will get the limiting value if the intensity is infinite. The law of proportionality has broken down because we have extrapolated it from a domain in which it is valid (low intensity) to a domain in which it is not valid (high intensity). The law of saturation, though more accurate over a greater range of intensities than the law of proportionality, will also break down at some intensity, for example, if we destroy the light meter by shining too much intensity on it.

The Ontological Argument most probably suffers from the same defect as the law of proportionality. First, it assumes that God is a being, whatever that means. Second, we have only a limited sample of beings, and most of those are not especially great. Yet the Ontological Argument attempts to extrapolate from finite beings (what is known) to infinite beings (what is not known). Once an extrapolation gets very far outside its range of validity, it can veer off in any direction whatsoever and probably will. That is why, for example, Newtonian physics cannot be extrapolated to high velocities, even though it is extremely accurate at low velocities. For the same kind of reason, the odds that the Ontological Argument is valid are vanishingly small.

* * *

Davies calls the Ontological Argument an attempt to define God into existence. He argues that it could be successful if it is augmented with extra assumptions, such as certain speculations about the existence of rational thought, but he does not say what those speculations are. His extra assumptions are tantamount to an *ad hoc* hypothesis, but he does not give enough information to make them testable. Alvin Plantinga, Professor of Philosophy at Notre Dame and, according to Pojman, one of our foremost philosophers of religion, has developed a complicated and generalized version of the Ontological Argument. [Pojman, 1987] Plantinga concludes that it is not irrational to believe in

God. If you recognize the distinction between irrational and nonrational, you will not consider that to be much of an advance. It is also a clear admission that the Ontological Argument does not work. Indeed, both Davies and Plantinga seem to me to be searching almost desperately for a way to make the argument work, whether it does or not. I cannot accept the Ontological Argument because (1) it is not logical, (2) it is circular, (3) it gives physical reality to infinity, (4) it is not testable, (5) it is an unjustified extrapolation, and (6) it is not supported by evidence.

The Argument from First Cause

This argument and several like it were developed by Thomas Aquinas (1225-1274 C.E.). Here is the Argument from First Cause:

> *There is a regular order of cause and effect (that is, cause precedes effect).*
> *Nothing can cause itself.*
> *It is not possible to project backward to infinity.*
> *If we remove a cause, the effect is lost.*
> *Therefore, if there is no first cause, there will be no effects.*
> *If we could project backward to infinity, there would be no first cause and similarly no effects.*
> *However, there are effects, so there must be a first cause.*
> *We call the first cause God.*

I think we can accept the idea of cause and effect; even quantum mechanics is causal in the sense that cause precedes effect. Nevertheless, it is an assumption, and there is no necessary reason why the universe must be causal. Our experience, however, suggests that the assumption is correct. Indeed, science itself proceeds on that assumption.

197

One problem with the Argument from First Cause is the statement that it is not possible to project backward to infinity. I assume that, by projecting backward to infinity, Thomas meant projecting backward to an infinitely long time ago.

An alternate way to look at projection back to infinity is to assume that there could not have been an infinitely large number of intermediate causes. In Thomas's time, they would have assumed that you cannot project backward through an infinite number of actions. We can, however, work our way through this alternate approach and show that the assumption is not necessarily valid. Indeed, it is reminiscent of Zeno's paradox: A runner cannot reach a goal because he first must travel halfway toward the goal, then half the remaining distance, then half the remaining distance, then half the remaining distance, and so on, an infinite number of times. Zeno thought it was impossible to carry out an infinite number of operations in a finite time and concluded that motion was an illusion. In fact, if the runner runs at constant velocity and takes one second to run halfway toward the wall, he takes one-half second to run half the remaining distance, then one-quarter second to run half the remaining distance, then one-eighth second, and so on. That is, the time required to get to the wall is the sum of the infinite series $1 + 1/2 + 1/4 + 1/8 + 1/16 + \dots$ seconds.

Even though there are an infinite number of terms in this series, the sum is finite and precisely equal to 2 seconds. Thus, motion is possible and, more importantly for us, it is possible to complete an infinite number of operations, even in a finite time.

The hidden assumption in this resolution to Zeno's paradox is that motion evolves continuously. If we similarly regard the universe as a continuously evolving entity, then we can clearly accept an infinite number of causes, as long as at least some of the causes are infinitesimal. If we can permit an infinite number of intermediate causes, then it is not obvious why we cannot permit an infinite time, even though at first glance doing so seems counter-intuitive.

F. C. Copleston, the Jesuit priest who was noted for his debate with Bertrand Russell, interprets Thomas's argument not to mean a time series of causes but rather a hierarchy of causes here and now. [Hick, 1964] Suppose, for example, that my boat is washed away by a wave. The wind is the immediate cause of the wave. Today, however, the tide is high, and that is a secondary cause that allows the wave to get to my boat. The immediate cause of the tide is the gravitational attraction of the sun and the moon. In addition, the tide is exceptionally high because the atmospheric pressure is exceptionally low, as it often is during a hurricane. The low pressure causes a partial vacuum that sucks the ocean water well above its normal elevation. Thus we have a hierarchy of causes, all in the present, that lead to my boat's being washed away. In Copleston's version of Thomas's argument, we cannot simultaneously have an infinite number of intermediate causes; therefore, there has to be an ultimate cause, and that ultimate cause is God.

But not all of the causes of the present situation are in the present. Why is the atmospheric pressure low today? Because we are experiencing a hurricane, which had its origin in a tropical storm several days ago. Thomas's argument ultimately boils down to a temporal series or several temporal series: in this case, a temporal series of causes that brought the hurricane, a temporal series of causes that brought the wave, and a temporal series of causes that brought about the high tide.

Copleston's version of the argument might have made sense when people believed, with Aristotle, that the natural condition of an object was at rest and that motion therefore had to be sustained somehow. If you accept Aristotle's argument and yet observe motion all around you, you may be inclined to look for a prime mover that is active at every instant in sustaining this motion. Today, however, we have replaced Aristotle's view with the principle of inertia: a body in motion remains in motion unless something actively stops it. We do not understand why mass has inertia, but neither are change and motion anomalies to us, and most of us would find it hard to argue that God is directly

199

responsible for the motion of each particle in the universe at every instant (but see [Swinburne, 1996] and the Argument from Design, below). It is therefore no mystery that the tropical storm develops into a hurricane, that it wends its way north, that it causes unusual tides, and so on. Copleston has not rescued the Argument from First Cause; at best, he has replaced it with a God-of-the-gaps argument (see below, this section).

* * *

At any rate, Thomas most likely intended his argument to refer to a temporal series of causes, because he says that an object cannot cause itself, for if it did so it would have to be prior to itself. He apparently could not conceive of a universe that was infinitely old. I can't either. But neither can I conceive of a beginning of time. What came before then? The idea of a beginning to time seems just as inconceivable as an infinitely old universe.

There is, however, empirical evidence that the universe has a finite age. It now seems likely that Thomas was right and the universe indeed had a beginning. Specifically, Big Bang theory argues that the universe began not quite as a point but rather, according to quantum theory, as a very small, very dense but finite blob. [Davies, 1992] Over the ages, this blob has expanded and become the universe as we know it.

Time is real to a theoretical physicist: as real as space. Davies argues that time came into existence during the Big Bang. Before that, there was no time. Thus, Big Bang theory resolves the dilemma and tentatively allows us to choose a finite universe over an infinitely old universe, a choice we cannot make by deduction alone.

You can argue with Big Bang theory, and there are a few problems with it. But, as Neil de Grasse Tyson [1996-1997] of the Hayden Planetarium and Princeton University notes, it is by far the most successful theory of the evolution of the universe. Still, it is difficult to imagine that it is accurate in detail, down to the last 10^{-33} cm (the diameter of the early universe, as quoted by Davies: a decimal point followed by 32 zeroes and then a 1). Our physics has never been tested on matter as dense as the early universe, and Big Bang theory extrapolates laws that work at lower densities to very much higher densities. It seems almost certain that some aspect of current physical theory will break down long before we reach the density of the Big Bang, much as the light meter saturated when the intensity got too high.

For a closer analogy, imagine a physical chemist who finds that the volume of a gas is directly proportional to its absolute temperature; that is, if you halve the absolute temperature, you halve the volume of the gas (as long as you keep the pressure constant). The physical chemist might reasonably conclude that the volume of the gas will be 0 when the temperature is 0 (absolute zero). The conclusion is incorrect, however, because the gas molecules have finite volume and form a relatively incompressible liquid at temperatures well above absolute zero. In the same way, I doubt that current physical theory will survive the extrapolation to unimaginably high densities (though it may have successfully described black holes, or collapsed stars with an extremely high density at their centers).

Still, it seems probable to me that the observable universe originated in a smallish explosion 10 or 20 billion years ago. The date is uncertain; until recently theories of the evolution of stars disagreed with cosmological theories by a factor of 2, though they are rapidly converging to the same age. The agreement is actually pretty good considering the difficulty of the measurements and the calculations. By contrast, the creationists' estimate that the universe is 10,000 years old is probably in error by a factor of 1 million.

Davies further gives arguments to suggest that the creation event was not a rebound from a previous Big Crunch, as some have proposed. His argument is convincing but depends on certain concepts, including entropy and relativity, being valid right down to the Big Crunch. At any rate, Davies rules out a cyclical universe and argues that the Big Bang can happen only once. Whether the universe will end up in a Big Crunch or whether it will expand forever remains an open question whose answer will depend in part on the resolution of the missing mass problem I discussed above (see "Falsifiability," Chapter 2).

Finally, we cannot rule out the possibility that our universe is one of many in what I will have to call a meta-universe (see also the discussion of the Anthropic Principle, below). For an analogy, imagine a weather map. It consists of high and low pressure regions. These regions rotate in opposite directions and may remain stable for days. They are called convection cells and are analogous to the upwelling regions in a pot of water that has been brought to a rolling boil. The air inside one cell mostly remains in that cell, and a molecule deep inside one cell will not "know" of the existence of another cell. It is possible that our universe is just a convection cell in a universe that is infinite in volume as well as infinitely old. There is no empirical evidence for or against such a model. In addition, it is possible that our universe was born as a black hole in an earlier universe and that the meta-universe is after all infinitely old. [Folger, 1997]

* * *

Let us agree, nevertheless, that the Big Bang model is approximately correct, and that Thomas was right in arguing that our universe has a finite age. That is, let us agree that the universe began at a roughly fixed time (roughly because quantum mechanics will not allow us to pinpoint the time precisely). Presumably, that event is the first cause that Thomas sought. That we do not understand the Big Bang is taken as evidence for a purposeful creator. Such an argument is an example of what is sometimes called *the God of the gaps*.

The science writer Kitty Ferguson has devoted the bulk of her book *The Fire in the Equations* [1995] to a God-of-the-gaps argument. She notes that there are many gaps in our knowledge. We do not know why the Big Bang happened; we do not know why the laws of physics are as they are; we do not know how life formed; and so on. We do not understand the human mind; we do not understand what Ferguson calls the supernatural nor what she calls meaning; we do not even know whether our scientific explanations are correct. Ferguson does not quite take these gaps in our knowledge as evidence for the existence of God but rather suggests that they leave open the possibility. Indeed, she admits that a God of the gaps leaves her on thin ice but argues that so does rejecting God because you think that science will some day explain everything. (She loves standoffs!) I find nothing wrong with her arguments, as long as she does not try to convince us that God must exist because we will never explain everything.

The gaps grow fewer and fewer, but some remain and probably always will remain. We have some idea how life began, but we do not understand the first step. Possibly we will never understand the Big Bang. We have reason to believe that our brains are in principle biological systems and that the supernatural is really caused by tricks of the imagination. Other gaps may be discovered, even as we understand more about our origins, and Ferguson suggests that complexity theory widens rather than narrows the gaps.

You could call Ferguson's argument "The Argument from Ignorance," though it is nothing like the Willful Ignorance of the Biblical literalists. To the contrary, Ferguson is curious, informed, and sophisticated. Still, she adduces no evidence that our lack of understanding is anything but lack of understanding. Rather, it seems to me, she tries valiantly but ultimately unsuccessfully to wring her conception of God out of our Ignorance. Neither Ferguson nor Davies presents a convincing argument that God as Big Bang is any more than an allegory, like Einstein's epigram, "God does not play dice." It does not aid in our understanding of the Big Bang and has no testable consequences.

* * *

The Argument from First Cause fails because (1) it provides neither reasoning nor evidence to suggest that the first cause was purposeful or, if you prefer, not reasonless. No less than the Ontological Argument, the Argument from First Cause (2) defines God into existence by establishing the probable existence of a first cause and then (3) confuses that first cause with the pre-existing notion of God. The Argument from First Cause is in essence (4) a God-of-the-gaps argument that assumes that anything we do not understand must have a supernatural explanation.

The Argument from Contingency

Philosophers define something as (logically) *necessary* if it can exist independently of anything else. Other things are called *contingent*. You and I, for example, are contingent beings in that, if our parents had never met, we would not exist. Saint Thomas thought that God was a necessary entity: [Pojman, 1987; Hick, 1964]

> *Some things exist for a time only and then go out of existence.*
>
> *If something is capable of not existing, then at some time it will not exist.*
>
> *If all things are capable of not existing, then there must have been a time when no things existed.*
>
> *But then, nothing would exist now, since there would have been nothing to bring it into existence.*
>
> *Things exist now, so they must have been created; that is, there must be something necessary in the universe.*
>
> *Whatever is necessarily existing was itself either caused by an outside source or not. Since we cannot have an infinite regression (the Argument*

204

> *from First Cause), there must be one thing that is inherently necessary.*
> *We call that God.*

The premise is wrong. Things do not exist for a time and then go out of existence. Things are collections of atoms, and the atoms, as far as we know, never cease to exist. They simply rearrange themselves (or else decay by fission into other species, but the matter still remains). The Argument from Contingency thus boils down to the question whether the origin and distribution of matter in the universe are contingent or necessary. That in turn boils down to the question whether the laws of nature are contingent. This is the question that Davies [1992] poses.

Davies agrees that it is difficult to see how there can be any necessary entities whatsoever. For one thing, says Davies, everything we know is contingent (or seems to be; if the evolution of the universe is strictly deterministic, then perhaps nothing is contingent. But let us continue with Davies's argument, even though he begs the question of contingency). According to the Big Bang theory, even the relative amount of each chemical element is related to the total mass in the universe. Davies speculates further that the values of the fundamental constants, such as the mass and charge of the electron and proton, may be connected to the total mass in the universe, that is, to the condition of the universe at the instant of the Big Bang. If that is so, then the only candidate for a necessary entity is the total mass, since it determines the laws and the fundamental constants. Otherwise, we would have to speculate whether the laws themselves are necessary.

By necessary, Davies means that the universe forms a closed system, and the laws are logically complete, so that everything follows from logical necessity. It is hard to see how that is possible, even in principle. First, there is the fundamental problem that no consistent logical system can be proved to be complete [Davies, 1992; Hofstadter, 1980]; this is known as

205

Gödel's theorem. In addition, we would have to be able to derive *from scratch* not only all the laws, but also all the fundamental constants and the total mass as well. Such a derivation is the aim of theoretical physicists who propose "Theories of Everything," that is, fundamental particle theories that are said to underlie all known physics. Davies says that, if we could derive these laws from scratch, then there would be no mystery, but I disagree: We still would not know where the universe came from, and we would still argue whether a necessary being necessarily created the Theory of Everything.

Davies finally gives up on that argument and postulates that the universe does not exist "reasonlessly." He labels the reason God and asks what properties this God would have. By assumption, he says, God would be rational. If we invoke an irrational god, we might as well settle for an irrational universe. Likewise, Davies's God is omnipotent and omniscient, as well as perfect, for he could not possibly have any defects. Such reasoning is no better than Kushner's rejection of God's omnipotence on esthetic grounds; the consequences (perfection) do not in any way follow from the postulate (rationality).

* * *

Saint Thomas was convinced of the existence of God, so he defined him into existence by "proving" the existence of a necessary being, much as the previous two arguments defined God into existence. But (1) the Argument from Contingency does not in any way suggest a purposeful or intelligent creator, and (2) Thomas's necessary entity need not be a being at all but rather could be the universe as a whole. That is not to say that the universe must exist, but rather that the universe or its total mass may be the best or only candidate for the role of a necessary entity. But its properties may be contingent on something we will never understand. It is still a leap to call that something God in the sense that I have defined it. I therefore cannot accept the Argument from Contingency.

206

The Argument from Design

The Argument from Design is most often presented in the way it was stated by William Paley (1743-1805), though David Hume [1990] earlier developed an argument equivalent to Paley's, specifically in order to refute it, according to Pojman. Paley's argument goes something like this: If I found a watch on the ground, I would naturally infer a watchmaker. I similarly infer a creator for the universe, since the universe has a purpose every bit as much as a watch has.

Assuming that the universe has a purpose is begging the question (assuming the answer). By purpose, I mean an overall purpose to the very existence of the universe itself and not the purpose we ourselves give to our lives. If God exists, then the universe may have such a purpose; otherwise it probably does not. Unless we know the purpose of the universe, we must not assume *a priori* that it has one.

Furthermore, a watch is obviously artificial. If I examine it closely, I will see tool marks. I may be able to figure out how it works and deduce its purpose, even if I have never seen one before. It may have an inscription such as "Made in Switzerland." I may be able to go to Switzerland, find a watchmaker, and verify the existence of watchmakers. None of these is true of the earth or the universe: We have never identified any tool marks, we do not know the purpose and have not deduced any purpose, and no one has ever found a 4.5 billion year old stone artifact (at the right geological stratum) with the words "Made by God."

The Oxford theologian Richard Swinburne [1996] presents an interesting variation on Paley's argument: He notes that all electrons have the same mass, charge, and spin, as do all protons, neutrons, and other particles. If he found a large number of identical coins in an archeological dig, Swinburne says, he would naturally infer a common origin and a common creator. The argument is no better than Paley's: The coins are obviously artificial, whereas the electrons are not obviously artificial.

207

Swinburne argues nevertheless that the "materialist" cannot explain why all electrons have, as far as we can tell, precisely the same properties. To Swinburne, the God hypothesis explains this otherwise puzzling peculiarity, whereas materialism is at a loss. I think, however, that Swinburne's argument may be able to deduce a common *origin* for all electrons, but he has no sound reason for inferring a purposeful *creator*. Most materialists will probably agree that all electrons have a common origin in the Big Bang.

Swinburne, by the way, argues that God is directly responsible for the motion of each particle in the universe at every instant (see Copleston's variation on the Argument from First Cause, this chapter). To Swinburne, God pervades the universe, is present at every point in the universe, and determines the trajectory of every particle in the universe at all times. God is omnipotent and knows everything exactly. Swinburne argues that the postulate of a single God who determines a multitude of laws and the masses of the fundamental particles is a simple or compact way of explaining the universe and its properties (see "Ockham's razor," Chapter 7). That is, he replaces several seemingly independent laws with a single God. In a way, this is a God-of-the-gaps argument, since it relies on our inability (so far) to construct a single theory that encompasses all the laws and predicts the masses of the fundamental particles. If we succeed in constructing such a theory, Swinburne's argument will be very much weakened. Swinburne, however, explicitly denies that his argument is a God-of-the gaps argument, but rather thinks that the God hypothesis more compactly explains the observed data.

Even to know the velocity of a single particle exactly requires an infinite memory, since it takes an infinite number of decimal places to state most numbers exactly. Yet God somehow knows the positions and velocities of every particle in a vast universe exactly—and knows them instantaneously. Where does God store the information, and how does he transmit it instantaneously from distant corners of the universe? Once he has the data, does he perform the necessary calculation instantaneously? Does he effect the necessary forces instantaneously? Or is he just a little bit ahead of us? If so, what does that say about the logical impossibility of predicting the future? (See "Reward and punishment," Chapter 3.) It takes very little reflection to see that Swinburne's hypothesis is not simple.

Yet, the Argument from Design is the most compelling of all the standard arguments. For if we can find evidence of either design or purpose with no *ad hoc* hypotheses and no assumptions such as arbitrarily equating the first cause with God, we will have a relatively sound argument that automatically suggests an intelligent or at least a purposeful creator.

Design seems to me to be a more important criterion than purpose. Something can be designed without purpose. Thus, if I happened upon a sand castle, I would deduce that it had a designer, even though it possibly had no purpose, or at least no lasting purpose, other than the amusement of the designer. By tacking purpose to design, Paley has made his task more difficult. After all, it is possible that God designed the universe with no lasting purpose.

Can we find design in the universe? Some people will see design in the growth of crystals or the symmetry of a sea shell. Others will find chaos in the rough and tumble, kill-or-be-killed way of nature. Some will find order in the solar system, where the planets revolve around the sun in comparatively stable orbits. Others will find chaos in the same solar system, where each planet and satellite seems as different from the others as a clam from an elephant. Some regard evolution as the mechanism whereby God carries out his purpose, whereas others argue that evolution gives the lie to the idea of intelligent design.

Let us examine two specific cases in which people have claimed to find evidence of design: evolution and mathematical physics. I will call these the Argument from Evolution and the Argument from Mathematical Physics. Let us take the Argument from Evolution first and then discuss an odd argument

209

called the *Anthropic Principle*, which (sort of) follows from evolution.

The Argument from Evolution

I have heard many people, including clergypersons who considered themselves Biblical literalists, argue that evolution was the mechanism whereby God created us. They usually agree (or admit) that the six days of Genesis are figurative, and some argue that we are perhaps not yet at the end of the sixth day. Their literalist reading of the Bible, incidentally, can be satisfied by Psalm 90:4, "In your sight, a thousand years are like yesterday." In other words, a thousand years to God is like a day to us. Thus, they assume that age of the earth is 10,000 of God's years, not our years. Since one of God's days equals 1000 years, one of God's years equals 365,000 years, so 10,000 of God's years equals 3.65×10^9 years, or about 4 billion years. This is in good agreement with scientific estimates of the age of the earth: 4.5 billion years. The age of the universe is more like 10 or 20 billion years.

The Argument from Evolution ignores several pertinent questions: Why did God go about the creation in such a roundabout manner? Why did he not create us and the universe as we are now? Was it easier to create a primitive universe and then guide it for 10 or 20 billion years? (To those who think that God is omnipotent and yet that he created the universe the easy way, I ask what "easy" and "hard" mean to an omnipotent being.)

* * *

The Argument from Design, as applied to human evolution, depends critically on the claim that evolution always proceeds toward increasing complexity: up the ladder of progress, if you will. In his book *Wonderful Life*, [1989] the paleontologist Stephen Jay Gould notes that evolution as the march of progress has been adopted as a given by our culture. Only the fittest survive, and the fittest are always perceived as stronger, faster, and more intelligent—never nastier, sneakier, and more ruthless, even though those are also attributes that can lead to survival. Hence, evolution has led directly and inevitably to us (or at least to intelligent life of some kind; Ferguson, [1995] the science writer, argues that God may have created the universe in the certain knowledge that intelligent life, though not necessarily human life, would evolve). This concept is so deeply embedded in the culture that I hardly need say more about it. Gould shows countless pictures of monkeys or apes gradually undergoing a transition to humans. Most are not serious, but they show that the ladder of progress is immediately understood by anyone at a glance.

Taxonomists recognize five kingdoms: animal, vegetable, fungus, and two kinds of single-celled organisms. Kingdoms are divided into phyla, phyla into classes, classes into orders, orders into families, families into genera (genuses), and genera into species. There are about 20 to 30 phyla today, according to Gould, depending on who is counting and classifying.

The conventional view is that evolution proceeds upward until all species are accounted for. That is, evolution is viewed not only as inevitable, but also as progress. Gould thinks it is neither. In an earlier article, "Life's Little Joke," [1991] Gould traces the etiology of the belief in the ladder of evolution. His subject is the descent of the modern horse. Textbooks show a direct line from the early ancestor Eohippus (now called Hyracotherium) through several intermediate stages to the genus Equus, which includes the horse. This is what Gould calls the ladder of progress: from the smaller and more primitive Eohippus to the modern horse.

Gould's argument is somewhat complex, but he shows the modern version, based mostly on careful interpretation of fossils called the Burgess shales. He shows evolution as a bush with many branches, not as a ladder. Each point where a branch divides into two twigs represents a new species. Many of the

211

twigs on the bush die; sometimes a whole branch, that is, a whole genus, dies. Specifically, the genus Equus is not a successful lineage but rather the last surviving genus of a dying lineage. Indeed, Gould claims that if we had looked at truly successful lineages such as rodents or insects, we would never have dreamed of drawing their descent as a ladder; it is simply too complicated. That is, we see a ladder of progress only in unsuccessful lineages, because only such lineages boil down to one or a few species. This is life's little joke (or perhaps Life's little joke): that *the model of the ladder could never have been derived unless we had looked at an unsuccessful lineage* such as that of the horse.

In an article in a special issue of *Scientific American*, Gould [1994] shows how increasing complexity can result from an undirected process and yet give the impression of direction. To illustrate Gould's point, I took a piece of graph paper and drew two axes, as in Figure 3. I labeled the horizontal axis "Complexity" and numbered it 1 to 10, and I labeled the vertical axis "Number of species" and numbered it 1 to 25. Then I made a mark at the point where complexity = 1 and number of species = 1. This represents a single species with minimal complexity: simple single-celled organisms, for example. We call it minimal complexity because it represents the simplest possible organism in the sense that anything less complex will not survive.

Next, I took a coin and labeled one side + and the other side −. I tossed the coin. If it came down +, then I moved up one unit in complexity and added a species. For example, if on my first move the coin came up +, then I added a species with complexity 2; if it came up + on my second move, then I added a species with complexity 3; if it came up − on my third move, then I added another species with complexity 2. There are now one species with complexity 1, two with complexity 2, and one with complexity 3. Species may evolve less complexity, by the way, but they never revert to an earlier form, so a step backward in my experiment means adding a species not returning to an earlier species.

212

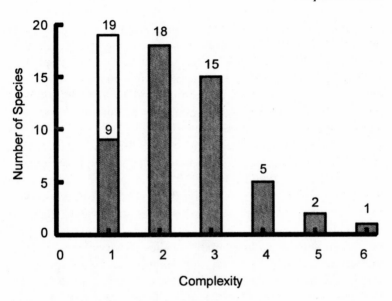

Figure 3. A graph of the number of species as a function of complexity after 60 coin tosses. As time (or the number of coin tosses) increases, a few species gain in complexity, but the number of less complex species also increases. The growth of complexity is a result of random drift, not of a directed process. The least complex species dominate.

I made one exception to the rule: If the coin came up – when the complexity was equal to 1, then I made no move, since by definition complexity cannot be less than 1. Gould calls this the "left wall," since you can graph complexity indefinitely far to the right (as long as you are willing to toss enough coins), but you cannot graph complexity farther left than the point where complexity = 1, since that point is, by definition, the minimum viable complexity.

I resisted the temptation to write a computer program but rather tossed the coin 60 times in all. Watching the points on the graph move left and right in a random fashion was instructive,

213

and I had to make many tosses before I reached a complexity of 4. The result of the simulation is shown in Figure 3. Already, there is one species with a complexity of 6. But the number of species at the low end of the scale has grown enormously. There are now nine species with complexity 1, 18 species with complexity 2, and 15 with complexity 3, but only two with complexity 5 and one with complexity 6. (There were also ten instances where the coin came up – when the complexity was 1; if I had considered those as standing pat, rather than ignored them, then there would now be 19 species with complexity 1.) This is about the way it is today; there are far more species of bacteria than of insects, more species of insects than of mammals.

When there are a great many species of varying complexity, the left wall is not especially important. The right edge of the graph will drift rightward whether or not there is a left wall. The left wall ensures only that the peak of the graph will also drift rightward. The drift of the peak may also contribute to what Gould would call the illusion of progress.

*   *   *

The model is obviously greatly oversimplified; in particular, it does not take into account extinction of species. That is why I could not subtract a species whenever I came up against the left wall but rather had to stand pat instead. I use the model only to illustrate Gould's point: A random or undirected process can give the appearance of direction toward increasing complexity, but the appearance is false and is the result of the left wall. Gould concludes that the appearance of direction in evolution is similarly false. Increasing complexity or direction in evolution is no more than random drift.

Research has backed up Gould's argument. [Oliwenstein, 1993] For example, the evolutionary biologist Dan McShea of the University of Michigan has examined the spinal columns of different vertebrates, or animals that have backbones, over a 30 million year period. He defined a complex spinal column as one whose individual vertebrae differ from each other in one of six different dimensions such as length or thickness. A simple spinal column is one in which each vertebra is more or less identical to all the others. He then asked whether the spinal columns of the fossil ancestors of animals such as squirrels and ruminants were more complex or less complex than those of modern squirrels and ruminants. If evolution was truly directed toward greater complexity, then the modern animals would have displayed more complex spinal columns than their 30 million year old ancestors. They did not: Sometimes the spinal columns became more complex, but as often, they became simpler. Other researchers, George Boyajian of the University of Pennsylvania and Tim Lutz of West Chester University, made similar observations of fossil shells called ammonoids. They discovered no tendency toward increasing complexity. McShea notes that increasing complexity is not necessarily a benefit; more-complex organisms have more things that can go wrong, for example.

That is not to say that complexity does not increase as the result of evolution; it does. But natural selection does not *force* increasing complexity. Rather, increasing complexity is the result of the gradual rightward drift of the *right* edge of Figure 3 as the result of random processes. In other words, my coin-tossing model was better than I had thought it to be.

* * *

Animal life was unicellular for nearly 3 billion years. Before that, there were some multicellular algae, but no multicellular animals. The first known multicellular animal life, or multicellular *fauna*, was the *Ediacaran fauna*, which appeared less than 700 million years ago. They were flat, two-dimensional creatures, some of whom looked somewhat like ferns. Gould writes that they probably died out, but might have been the ancestors of modern corals. More recent research has challenged this view, but they were extinct or nearly so early in

215

the great Cambrian explosion. This refers to a sudden explosion of animal life into countless new genera about 600 million years ago. A mass extinction followed the Cambrian period, possibly because an asteroid struck the earth. Asteroids strike the earth from time to time and are among the causes of mass extinctions. An asteroid almost certainly caused the extinction of the dinosaurs, for example.

The occurrence of mass extinctions leads us to the following model of evolution: During normal times, evolution occurs in a normal fashion. Never mind what "normal" means; it is unimportant whether gradually or episodically. But mass extinctions happen roughly every 26 million years. Sometimes well over half of all species can be wiped out [Kerr, 1995]. Which species survive may not be entirely random, but it is hard to see how any species can be adapted specifically to survive a catastrophe that happens so infrequently. Thus, we have to say that those species that survive, survive by pure, blind luck.

Gould [1989, p. 322] shows a drawing of Pikaia, which he calls the world's first known chordate. A chordate is an animal that has a backbone or a precursor of a backbone, and we are chordates. True vertebrates appear in the Ordovician period, immediately after the Cambrian, and certain intermediate stages appear in the late Cambrian. Gould does not claim that Pikaia is the ancestor of all modern vertebrates, but he notes that chordates were not especially common in the Cambrian period and suggests that, but for the extinction that ended the Cambrian period, the chordates might not have been equipped to survive the competition from other species and might have died out.

The dinosaurs were not unsuccessful and did not die out, as was once thought, because they had small brains and large bodies. They survived for 200 million years before they were wiped out by a mass extinction. Mammals had coexisted with dinosaurs for 100 million years, but they were small and insignificant. They did not challenge dinosaurs for supremacy but rather lived underfoot. We do not know why they survived the mass extinction, but they thrived only after the dinosaurs were removed.

Gould suggests that we go back in time, "rewind the tape" of evolution, and record it again. The new tape will almost certainly not resemble the old tape in any way. During normal times, evolution may progress in a way that could remind you of a grand plan. But the mass extinctions, though obviously determined by physical laws, are for practical purposes random occurrences that completely disrupt whatever grand plan may appear to be in effect. In the last 500 million years, there were at least a half-dozen mass extinctions in which 20 percent or more of all species were wiped out [Kerr, 1995]. Our own species came into existence only because an asteroid impact wiped out all of the dinosaurian competitors of certain earlier species from which we have descended.

Gould notes seven developments that were crucial to the development of human life. If any one of these had been absent, we would certainly not exist nor, possibly, would intelligent life of any kind. Very briefly, these are the development of a cell with a nucleus and other internal structures, the demise of the Ediacaran fauna, the Cambrian explosion, the rise of modern phyla after the Cambrian period, the rise of terrestrial vertebrates from a very specific and uncommon kind of fish, the rise of mammals after the destruction of the dinosaurs, and, finally, the development of an intelligent species of ape. Even the last step is problematic; our nearest cousins, Homo erectus, died out, and Gould does not believe that the Neanderthals would automatically have, as he puts it, taken up the torch if Homo sapiens had also died out. Only Homo sapiens is known to have developed counting or art; contemporaneous Neanderthals apparently did not. Start recording the tape again, Gould says, and Homo sapiens may die out in Africa.

Gould, it seems to me, ignores the inexorable rightward creep of the right edge. More important, however, he does not consider the possibility that intelligent life other than human might have developed in our stead. Why would not the

217

descendants of the dinosaurs have grown steadily more intelligent? Or, for that matter, why would not the early mammals have grown steadily more intelligent because of the competition from the dinosaurs?

Robert Wright [1990; see also Dennett, 1995] points out in a review of *Wonderful Life* that the Neanderthals might have gone extinct because we drove them to it. As Gould himself says, the earth has 5 billion years to go. It is hard to imagine that some other species, possibly the Neanderthals or their descendants, would not have become at least as intelligent as we, given enough time. And 5 billion years is apparently plenty of time; from the Ediacaran fauna to us required only 700 million years, or 0.7 billion years. Indeed, intelligent life has developed from scratch in well under 5 billion years (the present age of the earth). It is therefore hard to imagine that intelligent life would not relatively soon reappear if we disappeared tomorrow. Gould's argument that *we* would not have developed but for several contingencies is sound, but it does not follow that, therefore, intelligent life would not have developed at all.

* * *

Evolution may not be a directed process, but still the right edge moves inexorably toward increasing complexity. Is it possible that God started the universe in motion in the certain knowledge that 5 or 10 billion years (the lifetime of the solar system) are enough for intelligence to develop? Unless the universe is teeming with intelligence, he certainly did it on a grand scale. Just a few solar systems ought to have been enough. Such a postulate puts us or our proxy, intelligent life in general, back to the center of the universe, precisely the location from which Copernicus expelled us several hundred years ago. It is possible, but is it likely?

My answer is no. If God intended intelligent life to develop, then he should have been more careful and not allowed massive objects to collide with the earth with such regularity. The

218

Argument from Evolution may have been plausible before we learned of the periodic mass extinctions. But the survivors of mass extinctions cannot be predicted, even in principle, since it is impossible for any organism to have evolved resistance to something that happens only once every 26 million years. The evidence thus argues strongly against a plan, divine or other, that leads inexorably either to us in particular or to intelligent life in general. Mass extinctions give the lie to the idea that evolution was designed as a deliberate mechanism for creating intelligent life.

I regard the evidence from paleontology as nearly conclusive: Evolution is haphazard, particularly during a mass extinction, and therefore does not provide evidence for design or purpose. To the contrary, the existence of mass extinctions suggests precisely the opposite: that there is no design and therefore no purpose. For this reason, I reject the Argument from Evolution.

The Anthropic Principle

The Anthropic Principle is a form of the Argument from Design. In its "strong form," the Anthropic Principle says, in effect, that the universe is the way it is *in order that* intelligent life will evolve to observe it. Lee Smolin, [1997] Professor of Physics at Pennsylvania State University, calls the strong form an explicitly religious idea, which asserts that the laws are as they are because the universe was created by a god with intelligent life in mind. The assumption that the universe was intended to develop intelligent life seems oddly anthropocentric: Why not beetles? There are many more beetles than there are, say, mammals or apes, and many more species of beetles as well. Maybe the universe was created to see how many species of beetles would develop, and humans and their pesticides are merely a by-product, an unhappy accident from the point of view of the beetles or their creator.

219

The weak form of the Anthropic Principle is innocuous and simply asks why the universe is the age it is now. It answers, because that is how long intelligent life required to develop.

Promoters of the strong form, however, argue that the universe seems uncannily hospitable to life such as ours. In *Dreams of a Final Theory*, [1994] the theoretical physicist and Nobel Prize winner Steven Weinberg notes an argument that the chemical elements are formed in the interiors of stars, and most of the heavier elements could not have formed in abundance unless carbon had had an unusually high probability of being formed. Without carbon, there would presumably be no life, so the supporters of the strong form argue that carbon has such a high probability precisely so that life will develop. Countering this argument, however, Weinberg notes a calculation that shows that the properties of carbon could have been considerably different without significantly reducing the amount of carbon produced in stars. In addition, he points out that, if the physical constants in general were different, there may well have been different ways to produce the chemical elements.

Supporters of the strong form of the Anthropic Principle argue more generally that each of several fundamental constants has to be exactly as it is (or nearly so), or else life could not develop. The odds that every constant would have the right value are very small. You can make a similar probabilistic argument that the odds of my existing and having precisely my genes are also vanishingly small. The fact, however, is that I exist and have my genes and no one else's. The odds are 100 percent that I exist and have my genes. It is of no value to argue about the odds of my existing after the fact; I exist and that is that. I exist as the result of a myriad of occurrences, some probable and some not, but all of which actually happened.

My having my genes is not the result of a probabilistic process in this sense: Each gene has two or more forms such as the form for blue eyes and the form for brown eyes. If we made human beings by drawing these forms out of hats and later assembling the germ cells, then you could calculate the odds of assembling a human being with precisely my genes. Those odds would be very, very low. We do not, however, assemble human beings in quite that manner, so such a calculation is meaningless. I did not get my genes entirely by chance; I got them as a result of a continuing process in which chance played a part. But, for example, I had no chance whatsoever of inheriting the gene for red hair, since none of my recent ancestors had red hair.

The strong form of the Anthropic Principle confuses *a priori* probability with *a posteriori* probability, that is, confuses the probability of an event before it happens and after it happens. Assume, for example, that Babe Ruth has hit a home run that broke the windshield of my grandfather's car. The odds that the ball would have hit any given car were fairly small. There were a great many cars in the parking lot that day, so it was very probable that the car that the ball hit would not have been my grandfather's car. Finally, it is unlikely that the ball, if it hit my grandfather's car at all, would have hit the windshield. Therefore, according to the logic used by the supporters of the strong form, the ball could not possibly have broken my grandfather's windshield. But it did, and the odds that it did so were 100 percent.

Similarly, we have no reason to assume that the universe got its fundamental constants by drawing their values out of a hat, so we have no obvious justification for applying probability theory to the values of those constants. The universe is here and has the constants it has, with odds equal to 100 percent. It may have got those constants by a deterministic process that we do not yet understand; in that sense, the Anthropic Principle is a God-of-the-gaps argument. More specifically, there is no justification for assuming that the fundamental constants are independent of each other and for applying probability theory naïvely in the manner of the Boeing 747 argument (see "Taste," Chapter 2). Maybe the universe assembled itself in the way it did because of a series of causal relationships, just like the eye or the Boeing 747, and there is only one wholly independent constant, whose value could have taken any of a large range of values. Indeed, before the inflationary theory of the early universe, it appeared as if the total mass of the universe had to have a very precise value, but that argument has been superseded. [Lasky, 1999] Very possibly the other constants will be dealt with similarly.

Supporters of the strong form of the Anthropic Principle add that the universe is hospitable to life. In fact, we know of only one place where the universe is at all hospitable to life, and, if there are billions of stars with planets, it easy to see how life could develop somewhere by chance. Indeed, it almost has to happen—and probably has happened more than once. The Anthropic Principle seems to me to be a quintessentially circular argument, but a number of theorists have embraced it at one time or another, and Davies [1992] incorporates it with qualification into his own version of the Argument from Design.

* * *

Let us replay the tape of the last few paragraphs of the preceding section. The earth is 4.5 billion years old. About 3.5 billion years ago, as soon as the earth became cool enough for certain chemical compounds to exist, life arose. This much may be inevitable. For 2 billion years, however, the only life on earth was made up of certain simple cells, prokaryotes, that lacked a nucleus and other complex internal structure. Eukaryotic cells developed about 1.4 billion years ago. Some observers believe that the evolution of eukaryotes was necessary for the development of multicellular animals, but multicellular animals did not appear for another 700 million years. It took yet another 600 million years for intelligent life to develop.

Gould notes that the sun has only 5 billion years to go. Intelligent life took almost 5 billion years to develop (at least on earth). That is roughly half the lifetime of the earth. Suppose that it had taken two or three or four times as long. Suppose that eukaryotes had taken 4 or 5 or 10 billion years to develop instead of 2 billion. No one knows why eukaryotes developed, but the organelles (functional subsystems within the cell) are probably smaller, possibly parasitic cells that were engulfed by other prokaryotes and live symbiotically within them. [de Duve, 1996] Evidently such an occurrence was very chancy, or it would have happened sooner. Why not later? Would the universe not exist? Will it exist until the sun explodes, and then realize that there is no intelligent life and cease to exist? Will it cease to exist when there is no longer intelligent life?

Smolin [1997] lists a number of apparent coincidences, besides the formation of carbon, without which life would presumably be impossible. Smolin argues that life would be impossible without stars and, perhaps surprisingly, supernovas. Supernovas are necessary for the formation of heavier elements; the Big Bang formed practically nothing but hydrogen and helium. Hence, Smolin postulates that life will exist only in universes in which stars and supernovas develop.

Smolin notes that neither stars nor supernovas would be possible if certain fundamental constants had values that were different from the values we observe. He postulates that our universe is but one universe in a myriad of universes that are formed sequentially as black holes break away from already existing universes and themselves grow into fully fledged universes.

The universes are completely cut off from one another, so there is no way to detect them directly. This has led the journalist Patrick Glynn [1997] to scorn Smolin's hypothesis as science fiction because it involves only unobservable imaginary universes. This unfortunate attitude has led others to scorn religion because it involves only an unobservable imaginary god. Glynn leaves himself wide open to such an attack; his God is no less imaginary than Smolin's universes. Later, Glynn makes the same mistake as Prager and Telushkin (see "Wishful thinking," Chapter 3): dismissing Smolin's hypothesis because it leads a universe with no purpose other than the production of black holes.

More important, however, Smolin's hypothesis is falsifiable, whether or not the other universes are directly observable.

223

Specifically, Smolin suggests that the most common universe should be one that produces the most black holes. This is so because universes that are "good" at creating black holes will multiply faster than those that are not. Eventually most universes will be the black-hole producing variety. Thus, says Smolin, calculate the probability of producing black holes as you vary the values of the fundamental constants and see whether you get the fundamental constants we observe in our universe. His preliminary answer is "yes."

Smolin's hypothesis is falsifiable, whether or not the other universes are cut off from us. Specifically, if we perform all the necessary calculations and find that our universe is far from probable, then Smolin's hypothesis has been falsified, and the oddity of our universe's hospitality to life remains.

I have to agree with Glynn on one point, however: If anything sounds like science fiction writ large, it is the Anthropic Principle. It reminds me of a science fiction story about the man who thought that the universe was a figment of his imagination. Once he realized it, things started disappearing. The story ended when he asked, "What if I am a figment of my own imagi–?"

At any rate, assuming the Anthropic Principle is like Paley's assumption that the universe has a purpose and therefore must have had a designer. It is begging the question. It does not answer the question of why the universe is here but rather assumes that it is here to support intelligent life. I cannot accept it because (1) it is circular, (2) it is not falsifiable, (3) it is a God-of-the-gaps argument, and (4) it misuses probability theory.

The Argument from Mathematical Physics

Davies [1992] repeats an old question, "Why does mathematics describe the universe?" My first thought on hearing this question was, "No matter what we used to describe the universe, they would ask why it works." But the question is more interesting than that.

At least part of Davies's point is that mathematics that was developed for no particular reason or for another purpose often turns out useful in physics. For example, Werner Heisenberg developed his theory of quantum mechanics by writing tables of physical quantities and by inventing mathematical operations that allowed him to operate on these tables to calculate the energies of atoms in different quantum states [Weinberg, 1994; Gribbin, 1984]. What was odd about Heisenberg's operations was that they did not *commute*; that is, if you reversed the order of the operations, you got a different result. This is very unlike ordinary mathematics, in which, for example, 2 × 4 gives the same result as 4 × 2. According to John Gribbin, an astrophysicist turned writer, Max Born recognized Heisenberg's tables and operations as *matrix algebra*, a subject Born had studied years before but which was unknown to Heisenberg. Matrix algebra is only one example of a mathematical system that fortuitously turned out to be useful in physics. I suspect, however, that most of the mathematics that is actually used in science was, like calculus and statistics, developed for a specific purpose.

Davies further notes that many mathematicians are Platonists: They believe that the mathematical formalisms and theorems they derive exist independently and that they do not *invent* them but rather *discover* them. If that is so and if mathematics accurately describes the physical world, then, says Davies, perhaps mathematicians are discovering the works of God, that is, that God designed the universe according to mathematical principles.

The Platonist believes that any concept is a pale imitation of a real concept that exists only in the world of Ideas. If we see a chair, it is a facsimile of the Idea of a chair. The psychologist Rudolf Flesch, [1951] who was famous for a time as the author of the book *Why Johnny Can't Read*, discusses the formation of a concept and shows that concepts are taught, not discovered. He uses the example of a child who is introduced to an animal called a dog. Later she sees another dog, and later still another.

225

Eventually she learns what a dog is. Or does she? Flesch says no. She learns *what grownups call a dog*, which is by no means the same as learning what a dog is. To Flesch, "dog" is a concept that we invented, not a concept that we discovered. We could as easily have classified the animals we now call dogs in some other way; for example, if we had made no distinction between dogs and wolves, all the animals we today call dogs would be wolves. Dogs would exist, but they would be called wolves. The concept of "dog," however, would not exist, and there is, therefore, no reason to believe that the Idea of a dog exists either. That is, Ideas have no independent existence but rather represent concepts that we have invented.

Flesch illustrates his point with other examples. To take just one, we do not know when the concept "dog" was invented, but we know precisely when a chair was invented. Before 1490, people sat on stools or benches, but no one had ever sat on a chair that had legs and a back and was designed for a single person. The first known chair resides in the Strozzi palace in Florence and, according to Flesch, is an uncomfortable-looking chair with a straight slat for a back. Before that chair or one much like it was invented, the very concept of "chair" did not exist. The concept of a chair was invented because it was needed to describe a one-person stool with a back. Unless the world of Ideas can anticipate all possible inventions, the Idea of a chair did not exist until someone thought of building one and, even then, probably not until someone thought of naming it and distinguishing it from a bench. As the philosopher David Hume noted, all ideas are copied from real objects, not the other way around.

The mathematician Leopold Kronecker sounded more like Flesch than like Davies when he wrote, "God himself made the whole numbers—everything else is the work of men." The whole numbers 0, 1, 2, 3,... are the positive integers and 0. Kronecker seems to be saying that any civilization would have developed a system of counting, but the rest of mathematics has been invented by humans, rather than given by God. Einstein

226

[1954] went farther: "... the series of integers is obviously an invention of the human mind, a self-created tool which simplifies the ordering of certain sensory experiences."

In this view, the statement $7 + 5 = 12$ is not a universal truth or a logical necessity but is no more and no less than a shorthand that humans invented for the prescription, "Take 7 objects and set them aside; take 5 more objects and set them also aside; merge the two groups of objects; count the new group; you will find that you have 12 objects in all."

I would liken mathematics to chess (or any other game with rigid and specific rules). We invented the rules of chess, and those rules permit certain moves that lead to certain configurations of the pieces. Not all configurations are allowed. Now consider the configuration at the end of a certain game. That configuration arose because the players followed the rules and moved the pieces in a certain way. We may be arguing semantics, but did the configuration of the pieces come into existence the instant we defined the rules of chess, or did it come into existence when the last player made (or at least thought of) his last move? To the Platonist, the configuration existed before the game began, and the players merely discovered it. I would argue the contrary: At most, the *potential* for that configuration came into existence when we defined the rules of chess; the configuration itself did not exist until the player moved his last piece, any more than a chair existed until someone actually made one. If you prefer, that configuration did not exist until it existed. The players did not have to end up with that configuration or any other specific configuration but rather finished the game the way they did because of the moves they had made previously. They might have ended with another configuration, or they might have got to that configuration by a different set of moves.

So, I think, it is with mathematical formalisms. We invented the rules, often, like the rules of plane geometry, specifically to describe what we saw around us. If we lived on a far smaller sphere than the earth, a sphere so small that it did not look flat,

227

we might never have invented plane geometry but rather would have invented spherical geometry instead. The theorems are not "out there" waiting to be discovered any more than chairs or configurations of chess pieces are "out there" waiting to be discovered.

We use the expression to *derive* a theorem. Perhaps that is because a theorem is neither invented nor discovered, but rather falls into a third class. Theorems may be inevitable in the sense that you can derive only certain theorems and cannot derive both a theorem and its opposite, but that does not convince me that theorems exist before they are derived, any more than a concept such as a chair or a specific endgame exists before it is created. Rather, the theorems are derived subject to whatever rules we choose to apply. No theorem exists until the rules have been stated and the theorem has been derived. Neither is it especially surprising that sometimes two mathematicians derive the same theorem by different means; they all start with the same rules and often with similar training.

Still, the Argument from Mathematical Physics comes a lot closer to the mark than the Argument from Evolution, which (in light of the evidence relating to mass extinctions) is very probably wrong. The heart of Davies's argument is that, at the most fundamental level, the rules are simple. Whereas we used to have separate disciplines such as biology, chemistry, and physics, we now know that biology can be explained by chemistry, and chemistry by physics. A great deal of physics, such as gas theory, can be explained by atomic physics, and atomic physics can be explained by particle physics. And it now appears as if particle physics can be explained by interactions among a handful of fundamental particles known as quarks. There are still problems, but it is at least arguable that we may be approaching what Weinberg calls a final theory.

But is it the most fundamental theory, the underlying theory? Or is it simply as far as we can go because we cannot build bigger particle accelerators? Or because our theories and measurements are statistical? Or because the mathematics will become intractable? It is entirely possible that mathematics does not describe the universe, but rather describes only as much of the universe as we can understand. In his popular television series *The Ascent of Man*, the scientist and philosopher Jacob Bronowski [1973] notes in passing (pp. 164-5) that astronomy was developed early and elaborately and became the archetype for many of the physical sciences. Why astronomy? Why not medicine or biology or agronomy, which ought to have been much more practical sciences to develop? Bronowski suggests that astronomy developed because it was regular and showed repeating patterns: the first science that we could describe quantitatively, that is, by mathematics. Possibly mathematics describes only those parts of the universe that it describes, and we have been careful in choosing our fields. We have been successful so far because certain approximations have worked for us, but now we may be reaching the point where we cannot develop more-fundamental theories because we will not be able to solve the equations.

At one time, it looked as if there were 92 chemical elements. These were our atoms, and they were considered indestructible. Then Rutherford probed some atoms with a beam of electrons, and ultimately we found that the atoms were made up of electrons, protons, and neutrons: three fundamental particles. The laws were simple, and any physicist might have said that this mathematical simplicity was evidence for an intelligent creator. Before that, the same physicist might have based the same argument on Newton's laws.

Unfortunately, as physicists developed more-powerful accelerators with which to probe the subatomic particles, they discovered that the supposedly fundamental particles themselves could be broken into what was called a "zoo" of about 200 particles. The mathematical simplicity of three fundamental particles was gone. Our physicist would not have been able to point to a handful of particles and simple laws.

Today, it appears as if those 200 particles can be described in terms of a handful of particles called quarks, as well as the electron and its relatives, the photon, and the neutrinos. Weinberg calls the theory beautiful but notes that beauty is a learned concept; earlier physicists might have regarded our present stress on mathematical symmetry as numerology.

In 1894, the American physicist Albert Michelson claimed that the "grand underlying principles have been firmly established," and Lord Kelvin is said to have believed that there was nothing new to be done in physics except to make more- and more-precise measurements [Weinberg, 1994]. Yet, at that very time, other physicists were wrestling with the *ultraviolet catastrophe*, that is, the difficulty that the theory of radiation from a hot body predicted infinite power at short wavelengths. Barely six years later, in 1900, Max Planck announced his quantum theory of heat radiation. In 1905, Einstein announced his Theory of Relativity, his quantum theory of light, and his atomic theory of Brownian motion. Michelson's grand underlying principles were firmly displaced by deeper principles.

* * *

The moon and the sun look about the same size when we view them from earth. Therefore, when the moon blocks the sun, during a total solar eclipse, the sun displays a brilliant corona. It is extremely unlikely that the moon would have precisely the right orbit and diameter to block the sun and reveal the corona. If the moon were much closer to the earth, it would block the corona, whereas, if it were much farther from the earth, it would not block enough of the sun to allow the dimmer corona to be visible. The astronomers Michael Mendillo and Richard Hart [1974] of Boston University have used these facts to suggest that the earth-moon-sun system was therefore designed by a creator. The moon, however, has not always been the same distance from the earth. In the past, it was closer to the earth and wholly covered the sun and the corona during an eclipse, whereas in the future, it will be farther from the earth and will not wholly block the sun's disk during a total eclipse. The appearance of the corona during an eclipse of the sun can be used as an argument for a creator only during a specific era. Indeed, solar eclipses were not total until 280 million years ago; did God not exist before then? [Wicher, 1974] It looks suspiciously as if a mere coincidence is being taken as evidence.

Is Davies falling into the same trap? When Newton's laws seemed to be fundamental, you could have used the apparent order and simplicity of those laws as the basis of a theological argument. More to the point, when there seemed to be only three fundamental particles, you could have used that simplicity as the basis for your argument. Later, when there was a zoo of 200 particles, your argument would have broken down. Still later, your argument would have been resuscitated by the discovery of quarks. Suppose that we manage to smash quarks and reveal yet another particle zoo. It looks suspiciously as if our longing for a fundamental theory is being taken as evidence for a creator.

In addition, Weinberg's "final" theory has problems. In particular, it has so far proved impossible to relate relativity to particle theory. Further, the theory predicts that the proton decays, with a half-life of about 10^{30} years, into a positron and a neutral particle. (The present age of the universe, for comparison, is roughly 1 or 2×10^{10} years.) Efforts to measure the half-life of the proton have been unsuccessful; it evidently has a half-life at least 1000 times longer than the theory predicts. Far from having a final theory, we may well be up against a new incarnation of the ultraviolet catastrophe.

Davies and Weinberg, perhaps somewhat wishfully, think that they have nearly hit upon the underlying theory. Davies sees the hand of God in the relative simplicity of the theory. Weinberg [p. 244 and thereafter] sees the opposite. He argues that every step in the history of science has, rather, resulted in a "chilling impersonality" of the laws of nature: Copernicus and Galileo showed that the earth is not the center of the universe; Giordano Bruno correctly guessed that the sun was only one among many stars; and Edwin Hubble showed that our galaxy was only one among many galaxies. Similarly, modern biology

231

has demystified life by suggesting that it can be created by natural processes. And, as we have seen above, evolution is not a directed process.

* * *

I do not find the Argument from Mathematical Physics convincing. The fact that there are regularities and that we can find mathematical analogies for some of them is not surprising. Inferring a creator from these analogies is like inferring the existence of a physically necessary being from the existence of logically necessary propositions. The Argument from Contingency fails because that inference is not justified. The Argument from Mathematical Physics suffers from a similar defect.

Much, if not most, of the mathematics that is used to describe physical processes was designed specifically for that purpose. It works, in part, because the universe is ordered in a certain way, and mathematics is an orderly discipline. Thus, the question, "Why does mathematics describe the universe?" boils down to "Why is the universe orderly?" We do not know why the universe is orderly, nor do we know just how orderly it really is. At best, the order exists only at the deepest levels of reality, and we may never be able to discern that order because of the uncertainty principle. We could postulate a creator to explain that order if we were so inclined, but it is hard to see how the existence of order necessarily implies such a creator. The Argument from Mathematical Physics thus provides little or no evidence for a designer.

Mysticism and the Argument from Religious Experience

This is perhaps the most difficult argument to deal with, for mystical or other religious experiences seem objectively real to those who experience them. In *Quantum Questions*, Ken Wilber, [1985] a prolific author who is described on the book jacket as a scientist, argues that all mystics perceive substantially the same thing, which he calls the Great Chain of Being. Chain might not be the best metaphor, since each link in the chain subsumes all the previous links. Indeed, he draws the Great Chain of Being as a series of concentric circles (p. 15). The innermost circle is labeled "matter." The next circle is "life" and is taken to include living and nonliving matter. The next circle is "mind" and includes life and matter, as well as mind; next come "soul," "spirit," and "Spirit." With a lowercase *s*, spirit means the highest level to which we can aspire. Spirit with an uppercase *s*, however, is not enclosed within a circle but rather is represented by the entire page. It is what Wilber calls the Ground or Reality of all levels. Davies [1992, p. 226] notes that mystics often speak of being at one with God or with the universe. Sometimes they perceive a very powerful influence and claim that they have grasped what Davies calls an ultimate reality. In Wilber's terms, they have perceived the Spirit. It is probably fair to say that Western religions call Wilber's Spirit, God.

Wilber argues that mysticism can be put onto a scientific basis: Mystical experiences are replicable in that you can train anyone to be a mystic (though I assume that some are better at it than others), and all trained people have much the same experiences.

According to Wallace Matson, Professor of Philosophy at Berkeley, mystics generally agree that they have seen or experienced a higher form of reality than they can perceive with their senses. [Pojman, 1987, p. 120] They further see that all of reality is somehow unified; indeed, the mystic becomes so tightly entwined with this higher reality that he does not simply experience it but rather becomes an integral part of it. In addition, the reality is somehow "perfect"; evil and misfortune are mere illusions (!). Matson claims that optimism is therefore nearly universal among mystics. Finally, Matson notes that the mystic considers his or her soul to be akin to this reality, and he interprets the mystical experience as evidence that the soul can be separated from the body. In the West, the experience is taken to be a taste of Heaven.

I have no reason to doubt that anyone or nearly anyone can be trained to be a mystic, nor that the experiences are, in some ways, similar. These claims, even if true, by no means prove that the interpretations of those experiences are real, however. For example, schizophrenics report hearing voices. These voices seem real to them: just as real as the Spirit seems to the mystic. No one (except perhaps other schizophrenics) believes that real voices are really talking to the schizophrenics. It seems certain, rather, that these voices are the result of certain biological processes going on inside their brains. Nevertheless, these voices sometimes convince schizophrenics to believe outlandish conspiracy theories that no sane person would believe for one minute. The brains of mystics are also similar to one another; if you train them to generate a certain brain state, they will naturally report similar experiences.

* * *

People who are not mystics also occasionally report mystical experiences. Prominent among these today are *near-death experiences*. These are memories of either hallucinations or real visions experienced by people who have come extremely close to death but were revived. The memories are very vivid, and this fact is often presented as evidence that they are real, not hallucinations. But a hallucination is by definition something that fools you into thinking it is real, and hallucinations are well known phenomena. Indeed, vivid hallucinations can be induced by hypnosis, and they may be accompanied by observable changes in the brain state of the subject. [Concar, 1998] Therefore the vividness of the memories is no evidence whatsoever.

The near-death experience often, but not always, includes the feeling that you are leaving your body, or having an *out-of-body experience*. [Ebbern, Mulligan, and Beyerstein, 1996]

234

Frequently, you feel as if you are looking down at yourself on the operating table. Some patients report a *life review*, the feeling that their whole life has flashed before their eyes. Others report that they are falling or being propelled through a tunnel toward a bright light, or that they see a significant religious leader, a view of Heaven, or a deceased relative. They interpret the visions as evidence that they were on the verge of the afterlife. Children, by contrast, rarely experience the death of a friend, and children often report seeing living friends; this fact suggests that the children's experiences, at least, are hallucinations. [Blackmore, 1991]

Near-death experiences may be more common now than in the past owing to the large number of people who, in the past, would have died but are today resuscitated by modern medicine. Many of these patients report that they have died and returned. Whether or not their visions are real, however, they have, by definition, not died but rather stopped breathing and were resuscitated.

The psychologist Susan Blackmore [1986] had a drug-induced out-of-body experience and found herself floating over Oxford. Later, she hoped to have one when she underwent surgery but did not. Blackmore is a keen observer and was seeking evidence for psychic powers, though she later gave up on the concept. Nevertheless, she reports that her out-of-body experience could not have been real, because she had got the map of Oxford wrong and, more important, all those red roofs she saw in her vision were in reality gray. Further, Blackmore thought she saw herself during her experience. When she returned to reality, she asked herself pertinent questions: How can you get out of your body and see yourself, if your eyes are still firmly planted in your skull? If you can get out of your body and transmit physical information to your (physical) brain, then there must be some mechanism, and your out-of-body self ought to be detectable. But no one has ever detected, as Blackmore puts it, floating eyes or any other physical evidence for an out-of-body experience.

A British psychiatrist tried to test the reality of out-of-body experiences by putting pieces of papers with numbers out of sight in the operating room of a hospital. [Ebbern, Mulligan, and Beyerstein, 1996] So far, no one has correctly recited the numbers. Another investigator tried a similar experiment with subjects who claimed that they could have out-of-body experiences at will. [Blackmore, 1986]. One subject succeeded, but she was not watched or monitored on videotape. Her brain-wave pattern, however, suggested that she had climbed up and read the number. The experiment, like most others in this field, is at best inconclusive.

Blackmore [1991] presents a very simple physiological model for the appearance of the bright light. She notes that our visual acuity is best in the center of the visual field and that therefore many more neurons in the visual cortex (the part of the brain devoted to vision) correspond to the center of the visual field than to the periphery.

When a light-sensitive cell in the retina is exposed to light, the information is carried along a neuron to the brain, where another neuron becomes activated, or *fires*. Most neurons are in a quiescent state most of the time, owing to a process known as *inhibition*. That is, some neurons inhibit others, or prevent them from firing. Neurons normally fire in response to a stimulus.

When the brain is deprived of oxygen, however, neurons begin to die. As they do so, the inhibition necessary for proper functioning is relaxed. In consequence, neurons become *disinhibited* and start firing at random but with increasing frequency as inhibiting neurons fail to function or die. When a neuron in the visual cortex fires, you experience it as a flash of light, precisely as if the retina had been excited; you cannot tell the difference. But most of the neurons in the visual cortex correspond to the center of the visual field. At the beginning, therefore, you see light mostly in the center of the field. As neurons become disinhibited in increasing numbers, the intensity of the light in the center of the visual field increases. Light spreads from the center of the field to the periphery—light at the end of the tunnel.

Visions may be explained similarly: The neurons in a region related to a visual image, for example, may become disinhibited, and you think you see a long lost relative. Thus, Blackmore argues, we all see visions or light at the end of the tunnel because we all have the same shutdown mechanism.

Why do the visions seem so realistic? Because you cannot tell the difference between a real experience (which causes neurons to fire) and an internal state that causes the same neurons to fire. This is especially so when your brain is shutting down and you are not aware of external stimuli, which would otherwise give you a handle on reality. I will further conjecture that the floating feeling that is often associated with a near-death experience results from the interruption of stimuli from tactile nerves that would otherwise tell you that you were lying in bed.

The journalist Patrick Glynn [1997] argues that near-death experiences are "intimations of immortality." He cites a study by the physician Michael Sabom, in which Sabom isolated six patients out of a set of seventy who had had out-of-body experiences. These six described their operations and included details that they "could not possibly have observed from their physical vantage point." The six cases are less than ten percent of Sabom's sample, and you could accuse him of dredging for data in the manner of Michael Drosnin and other numerologists (see "Signs," Chapter 3). The argument, however, is not as pertinent here, because only a single verified instance of an out-of-body experience is enough to demonstrate their existence.

Nevertheless, it is important not to be fooled by your preconceptions of what a patient could or could not have known. A man who sees his wife and son arriving at the hospital could be merely choosing from a small set of possibilities. A woman who remembers being hit on the back of the head may well be confabulating from facts she gleaned after her recovery (why would she have had the out-of-body experience *before* the blow was struck?). The psychologist Jean Piaget, for example, had

237

vivid memories of being kidnapped as a small child. When he was a teenager, he learned that the event had never truly happened. Having heard of the incident many times, the young Piaget had simply confabulated memories of it. Similarly, when Ebbern and his colleagues investigated an out-of-body experience that had been touted as a strong case, they found that a certain sneaker that the patient could not have known about was in fact plainly visible from the street. Such visions are weak evidence and are not demonstrated to have statistical significance.

Visions of the operating room are more compelling. Nowadays, however, most patients are very well aware of what goes on in an operating room, since they have seen both real operations and actors simulating operations on television. It is certainly plausible that these six are the only patients among the seventy who happened by chance to describe their own situations right. (Recall our discussion of the number pi under the heading "Signs," Chapter 3. In any large random set of numbers, there will very probably be a series of digits that appear nonrandom but are not; similarly, in any large random set of incoherent recollections, there will very probably be some that appear coherent but are not. This is a point, by the way, that Glynn emphatically does not understand.) Sabom does not appear to have implanted "memories" into his patients' brains, and his assessment may be accurate, but the evidence is at best equivocal.

Glynn argues that near-death experiences are different in kind from ordinary hallucinations caused by *anoxia*, or lack of oxygen in the brain, or other physiological processes. Specifically, he argues that anoxia causes a disruption of orderly brain processes, not coherent memories, and that out-of-body experiences do not correlate with known cases of anoxia. Additionally, he quotes a former Royal Air Force pilot who has both experienced anoxia and had a near-death experience on separate occasions. The pilot claims that there was no similarity. This anecdote points out precisely why it is difficult to take anecdotes seriously: There is no evidence that the pilot experienced anoxia severe enough to bring about symptoms similar to a near-death experience.

Further, Glynn is incorrect in describing either near-death experiences or the reaction to anoxia as if they were consistent from patient to patient. In fact, near-death experiences vary widely from patient to patient. Not all patients experience the life review, and not all see the bright light. Indeed, there is not a single experience that is common to all near-death experiences. [Blackmore, 1996]

Similarly, reactions to (for example) anoxia depend on the cause of the anoxia, its speed of onset, and its duration. Humans placed in large centrifuges to simulate the experiences of pilots have reported symptoms similar to near-death experiences, and two have reported out-of-body experiences, even though they were in no danger, as have people under the influence of psychoactive drugs.

Still, it seems to be true that people who are close to death more often report the bright light, an out-of-body experience, or the presence of friends or other beings. Many of these patients know or believe that they are dying; if only for that reason, a near-death experience is qualitatively different from hallucinations brought on by anoxia or other physiological stresses. Thus, there is probably a psychological element to near-death experiences, as well as a physiological element. Perhaps this psychological element is responsible for the life review.

A near-death experience is a complicated phenomenon, probably brought on by the confluence of several processes. Glynn argues that, because the cause is not anoxia, not hypercarbia (an excess of carbon dioxide in the brain), not the effect of endorphins (brain hormones that bind to opium receptors and may reduce pain or cause a feeling of well-being), not the effect of exciting the right temporal lobe (which is thought to be the part of the brain responsible for emotion), near-death experiences must be objectively real. Such thinking is too simplistic. Each symptom can be stimulated by one or more of those factors. The near-death experience can therefore be explained as a purely physiological effect brought on by a combination of Glynn's and other factors, probably combined with the belief that death is imminent.

Thus, though near-death experiences are interesting, they are wholly unacceptable as consistent evidence for an afterlife or another plane of existence.

* * *

One surprising thing about near-death experiences is that they are unsurprising. No patient ever "returns" with a report that conflicts with anything in his or her worldview. No one has ever returned with a daring prediction, for example, that was later verified. Neither has any mystic. The reports of mystics seem to be culturally determined, in the sense that Christians view Heaven, whereas Buddhists experience nirvana. In other words, when you have a mystical experience, you see or experience only concepts that are already in your brain. There is therefore no reason to believe that these perceptions are objectively real in the sense of describing something external to your brain.

I occasionally hear writers of fiction interviewed on the radio. Practically every one of them says very matter-of-factly that her characters tell her where the story will go next. Now, she does not believe that literally. Rather, she is describing how her thought process appears to her. She knows perfectly well that her characters are not real and do not talk to her; it merely feels that way. In fact, the plot unfolds in the author's brain and nowhere else. The same is arguably true of mystical visions and near-death experiences.

* * *

Examining case histories of people who claim to have been kidnapped by aliens may be instructive. [Blackmore, 1998; Nickell, 1998] Most are kidnapped while in bed. They report being paralyzed and then examined and sometimes sexually abused by aliens. Sometimes, they are subjected to grueling physical examinations, but there is never any objective evidence, such as bruises or contusions, of these examinations. The aliens usually look like the aliens from science fiction movies. In the Middle Ages, people reported being seduced or raped by evil spirits (incubi and succubi), rather than bug-eyed monsters. Paintings of such evil spirits look to me like the bas reliefs or gargoyles that were built into churches to keep away evil spirits. I suggest, therefore, that reports of alien encounters are, at a minimum, culturally biased. Indeed, the aliens' keen interest in the reproductive systems of humans also suggests that the source of the visions is internal rather than external and that the visions are no different from medieval visions of incubi and succubi. Why after all, would an alien, who is a different species from yours, care about your reproductive system?

At any rate, there is a simple alternative explanation. Certain kinds of dreams are well known to psychologists. They are called *hypnogogic* dreams and *hypnopompic* dreams. Hypnogogic dreams are dreams associated with the state of drowsiness or semi-consciousness that precedes sleep, whereas hypnopompic dreams are dreams associated with the same state but preceding awakening. The distinction is irrelevant to us, so let us call them simply "waking dreams." For an account written by a man who experienced waking dreams and was terrified of them before he found the explanation, see [Huston, 1992].

People who report waking dreams report that they are paralyzed during the dream. We are all paralyzed during some phases of sleep, so this in itself should be no surprise. They also report that the dreams are very realistic. Most people recognize them as dreams, but some apparently do not.

Given two explanations—that the person was having a well known kind of dream or was abused by aliens—and no objective

241

evidence for an alien encounter, I think we have to assume that he was dreaming. Reports of visions, as of the Virgin, also have to be taken lightly, in part because these visions are almost never anything a suggestible person might not simply imagine. According to Pojman, they are almost always specific to the culture of the visionary.

* * *

Pojman [chap. II.5] counters the claims of Wilber and Matson with the argument that there are too many varieties of religious experience to allow the kind of generalization that they make. He lists 15 documented religious experiences, all very different from the others, in a rough order from the mystical to the material. At one end of the spectrum is a mystical experience that I take to be similar to those described by Wilber; at the other end is an atheist who feels that the universe has manifested itself to her as a "deep sense of nothingness." Between these, Pojman lists, among others, a vision of the Virgin, Jesus appearing to Paul, Athena appearing to Achilles, a man told to execute all infidels, a man who feels the Trinity but cannot explain it, and Plato, who feels the presence of a creator (the demiurge) who is neither omnipotent nor omniscient. Far from being universal, these experiences could not be more different from one another. Besides Plato and the two atheists, three of the experiences do not include God at all, and, Pojman notes, certain branches of Buddhism and Hinduism do not involve God either, yet they are mystical religions.

One of Pojman's examples is Arthur Conan Doyle's daughter. Pojman describes her as a guilt-ridden woman who sensed the presence of her long-dead father, whom she had neglected in his old age. In her vision, he forgave her for this neglect. How, Pojman asks, does this mystical experience show that there is a God whom we should worship? Or does it suggest ancestor worship instead?

After reading Pojman's listing of mystical experiences, I realized that I had had such an experience myself but had never before considered it mystical. On being asked to conceive of no longer existing, I perceived myself falling or floating through a blackness blacker than any I had ever seen. I had no sensory stimulation whatsoever and no thoughts. I somehow "knew" that there was nothing in this universe, that the blackness extended to infinity, and that I would fall or float forever. If I had been more suggestible, I am sure I would have had a feeling of vertigo as well. I can reproduce the experience almost any time I want to, provided that it is quiet enough. Do I regard it as evidence that there is no afterlife? Not at all. I regard it as my imagination confirming what I already believe for other reasons.

That people have religious experiences is incontrovertible. The proof that these experiences support their beliefs in an objectively real God, however, is completely lacking. Pojman notes that the character of the experience is determined by the background, or core beliefs, of the observer: Christians, he says, see Jesus, Hindus see Krishna, Buddhists see Buddha, and nonbelievers see nothingness. How can we prove that these disparate visions converge on a single truth? We cannot.

The Oxford theologian Richard Swinburne [1996] (whom we have met in connection with "Reward and punishment," Chapter 3, and "The Argument from Design," Chapter 6) argues in this regard that if people claim to have experienced something, then we should believe them, provided that we have no reason for disbelieving them. He calls this precept the Principle of Credulity (an unfortunate choice of words, since "credulity" implies a willingness to believe on scant evidence). According to the Principle of Credulity, people are good observers and should be believed unless there is a sound reason not to believe them. To borrow one of Swinburne's examples, if you said you had read a page from a book, I would not disbelieve you, unless you said you had read it at a distance of 100 meters, in which case I would have ample reason to disbelieve you.

243

I have already discussed several reasons for not accepting people's recollections or hallucinations at face value (see "Anecdotal evidence," Chapter 2, and the discussion of near-death experiences, directly above), so I will belabor this point with only a single example. *Achromatopes* are people who are wholly unable to perceive color, either because they lack color-sensing *cones* in their retinas or because their cones are vestigial. [Sacks, 1996] Such patients can see only with their *rods*, which are the receptors that normal people use at night and which cannot detect color. Achromatopia is very different from the more common red-green colorblindness. What other people call "color" must be completely mysterious to achromatopes. Yet they presumably believe others when they claim an ability to see color. Swinburne argues that, similarly, we should believe those who claim to have experienced God.

Are the situations parallel? Clearly not. A colorblind person can easily perform an experiment to determine whether the normally sighted people are telling the truth: Just get several samples of colored materials such as swatches of paint and ask a great many people to identify the color. If they agree on the color and especially if their agreement transcends cultural and linguistic boundaries, then they have an ability that is foreign to the colorblind person.

We cannot perform a similar experiment on people who claim to have had religious experiences. If, nevertheless, we attempt a retrospective survey, we will presumably find, with Pojman, that their answers vary widely and are culturally determined. There is no more reason to believe that religious experiences are objectively real than to believe that any other internal reality is also objectively real. The Principle of Credulity may be applied to statements that can be verified objectively. It is, however, dangerous to apply it to statements that cannot be verified objectively. Any jilted lover could have told Swinburne as much.

* * *

When supporters of the Argument from Religious Experience are pressed for evidence that supports their visions or feelings, they often reply that there are "other ways" of knowing. This is another way of saying that they consider themselves intuitive thinkers, not deductive. John Polkinghorne, [1994] a physicist turned Anglican minister, talks of "the ability of understanding to outrun explanation" and relates it to religious experience. He cites two anecdotes of scientists, Henri Poincaré and Paul Dirac, who reported sudden flashes of insight into the solution to a problem they had been working on. Polkinghorne regards such flashes of insight not as unsubstantiated assertion but rather as an ability to grasp the heart of a matter in totality, without detailed analysis. He regards religious insights in the same manner. That is, he regards religious experiences as self-authenticating.

Carl Sagan [1977] agrees as to the importance of intuition and suggests that critical thinking alone will never lead to new ideas. He adds, however, that there is no way of knowing whether an idea arrived at intuitively is correct unless it is subjected to critical thinking. I referred earlier to Heisenberg's noncommuting mathematics (see "The Argument from Mathematical Physics," Chapter 6). Steven Weinberg, [1994] the Nobel Prize-winning theoretical physicist, says that he has tried but cannot understand Heisenberg's motivations for taking the mathematical steps in the paper in which Heisenberg describes his tables. Heisenberg had somehow gotten them by intuition. His mere intuition was not enough, however. His theory had to be proved by comparison with experiment, or it would not have been considered correct. By contrast, Einstein proposed a thought experiment that was designed to refute some of the assumptions of quantum mechanics. Einstein's experiment (actually a variation of the experiment he proposed) came out precisely as he had thought it would not. Relying on intuition is not enough, even for an Einstein.

245

Polkinghorne is well aware of the need for verification in science and is at pains to note that both Dirac and Poincaré later verified their insights. Suddenly compartmentalizing, however, he ignores the need for verification of *religious* insights, not to mention the myriad of scientists or laypersons who were not Diracs or Poincarés and whose insights turned out to be dead wrong.

In addition, Polkinghorne is a believing Christian and a supercessionist; that is, he believes that, with the birth of Jesus, God transferred his covenant from the Jews to the Christians. He believes that Christians should abhor anti-Semitism but nevertheless present the truth as they see it to their "Jewish brothers and sisters." We can fairly assume, therefore, that Polkinghorne would not accept as true the insight of a Jew who believes that God's covenant is still with the Jews. Polkinghorne has a *triple* standard: Scientific insights have to be verified; religious insights do not have to be verified; but some religious insights are not true. More specifically, the religious insights of Polkinghorne and those who agree with him are apparently self-authenticating, whereas other such insights are not. In short, Polkinghorne's case for the truth of religious experience is at best inconsistent and at worst disingenuous. The affirmation of the Oxford theologian and scientist Arthur Peacocke [1993] that non-Christian religions are also a path to the "reality" he calls "God" is not only not arrogant but also less likely to lead to religious intolerance.

* * *

Even Pojman, who lists the varied and incompatible kinds of mystical experience, seems to accept that a religious experience is proof enough for the person who has had the experience, though he grants that it is not possible to convince another who has not had the experience. I cannot agree with his premise. The argument is about equivalent to Ricky's statement, "If I believe it, then it is true" (see "Scientific method," Chapter 2).

Go back to the optical illusions (see "Internal reality," Chapter 2). Without a ruler, you would believe firmly that one line segment was longer than the other. No matter how firmly you believe it, however, the line segments are the same length, and your intuition has failed you. I do not know whether the experiences of mystics are objectively real and not figments of their imaginations. But there is no evidence that they are objectively real, and I consider other, simpler explanations at least equally probable. I therefore cannot accept the Argument from Religious Experience as serious evidence for a creator.

Conclusion

The Arguments of the philosophers fare no better than the popular religious beliefs of Chapter 3. The Ontological Argument is so badly flawed that even its supporters point to the need for supplementation. The Argument from First Cause assumes, possibly correctly, that the universe came into existence a finite time ago but fails to provide evidence that the first cause was purposeful. The Argument from Contingency assumes the existence of a necessary entity and then confuses that necessary entity with a purposeful entity. The Argument from Evolution fails because evolution is haphazard, and periodic mass extinctions make it more so. The Argument from Mathematical Physics may come closest to the mark, but it too fails because it begs the question why the universe appears to be orderly. Finally, the Argument from Religious Experience is unusually weak, since religious or mystical experiences cannot conclusively be distinguished from hallucinations.

In the next chapter, I will summarize my belief that, whether I like it or not, the universe is governed by impersonal laws and those laws reach right into our very brains and govern how we think and act.

Je n'ai pas besoin de cette hypothèse (I do not need that hypothesis).

PIERRE-SIMON LAPLACE

Chapter 7
Experimentalist's universe

Table 1 lists the arguments that relate to the existence of an intelligent or purposeful creator. It has two sections: what I have called popular arguments and philosophical arguments. The line is a bit blurry in that evil and altruism, for example, could just as well have been called philosophical arguments. The table is also a bit schizophrenic in the sense that, above the line, we were considering popular arguments which made no distinction between God the creator and God the active agent, whereas, below the line, we concerned ourselves only with the existence of a purposeful creator. As you have no doubt deduced by now, I consider that all of these arguments fail or provide only the weakest of evidence.

No argument can be conclusive. If we had found several arguments that had even a modest probability of proving the case, we might have said that we had built a strong circumstantial case. I argue the opposite: The case for an intelligent creator is so far from proved that it is almost nonexistent.

* * *

Charles Pellegrino, whom we met earlier in connection with the Exodus (see "Signs," Chapter 3), is a paleontologist turned space researcher turned archeologist. His book *Return to Sodom and Gomorrah* [1994] cannot be described in a few words, so let it suffice to say that he searches for evidence and dates regarding such Biblical tales as the myth of the Exodus. Before each expedition, Pellegrino receives a blessing from his priest, John

MacQuitty. Early in the book, Pellegrino says that he is an agnostic, and MacQuitty knows it full well. MacQuitty considers Pellegrino's questioning to be God's gift to Pellegrino.

Pellegrino's argument in favor of agnosticism is that you cannot ever know with certainty, and the day he stops questioning and becomes either an atheist or a believer, he will stop being a scientist. I agree that we do not have the evidence to decide with certainty and probably never will. But we cannot ever know anything with certainty. In that sense, we are all agnostics about everything, or at least we ought to be. That is, all our decisions and all our beliefs are necessarily based on incomplete evidence. For example, as I noted above, physicists believe in the Law of the Conservation of Energy because no exception has ever been found, even though we know we have not examined all possible cases. Most of us doubt that an exception will ever be found, however, and it is in that sense that we "believe" very firmly in that law. Nevertheless, no competent physicist will deny that we could be wrong, and our belief in conservation of energy is in that sense provisional. In the same way, I think it is appropriate to take a stand on the existence or nonexistence of a creator. Refusing is, to me, a little like refusing to vote because we cannot predict the winner. Taking a stand, however, is not irrational and does not mean that we are no longer scientists, provided that our stand is provisional and not dogmatic.

Ockham's razor

The English cleric William of Ockham (1270-1347) was a *nominalist* who took a position somewhat more extreme than that of Rudolf Flesch. That is, he regarded concepts such as "dog" or "horse" as mere constructs or conveniences with no reality whatsoever: a shorthand that allows us to convey our meaning without having to draw a picture of a horse. The extreme form of nominalism held that nothing exists except what you can perceive through your senses. In modern terms, we would call William an empiricist. It is hard to see how an extreme

250

Table 1. The arguments.

Argument	Type	Conclusion
Miracles	Anecdotal	Fails
Reward and punishment	Anecdotal	Fails
Morality	Empirical	Fails
Signs	Anecdotal or statistical	A few anomalies
Evil and misfortune	Empirical	Counter-evidence, if anything
Altruism	Empirical	Fails in light of sociobiology
Ontological	Deductive	Fails
First cause	Empirical	Fails
Contingency	Empirical	Fails
Design	Empirical	Fails as a general argument
Evolution	Empirical	Fails in light of mass extinctions
Anthropic Principle	Empirical	Circular
Mathematical Physics	Empirical	Weak evidence, at best
Religious Experience (mysticism)	Anecdotal	Fails

nominalist can believe in God, since God is an abstract concept, and William recognized this point. He abandoned any attempt to prove God's existence but chose to believe on the bases of faith and revelation. The Church, nevertheless, considered him an agnostic and tried him for heresy.

William is remembered primarily for a scientific or philosophical principle known as *Ockham's razor*: Plurality should not be assumed without necessity. That is, when there are competing theories or explanations, choose the theory that is simplest in the sense that it has the fewest entities or variables. As a corollary, you might add: Abandon the simplest theory only when it is proved unsuitable. A monotheist who accepted the Ontological Argument, for example, might use Ockham's razor to argue that there is only one infinite being, since that theory has the fewest entities. He would abandon that belief only in the face of evidence that there were, in fact, more than one such being.

How can we apply Ockham's razor to the search for a creator? We are faced with a universe. There are two choices: Either it exists for no reason (or at least no reason we are likely to discover) or it had a purposeful creator. Richard Swinburne, [1996] whom we have met several times, argues from Ockham's razor that one all-powerful creator is simpler than a multitude of elementary particles. There is, however, almost no credible evidence for an intelligent creator. If we assume a creator, we have to postulate an incredibly powerful being who had the capacity to create a universe with not merely billions of stars but rather billions of galaxies and quintillions of stars (there are even more galaxies in the universe than there are stars in our galaxy). Where did this being come from? Where does he live? Is he still here? What is his purpose? What does he want of us? Postulating a creator creates endless complexity. In addition, we have to ask where that creator came from. Who created him? Who created the creator of our creator? We are faced with an infinite regression. To assume a creator, therefore, multiplies entities unnecessarily.

I therefore postulate

The universe is not the product of an intelligent or purposeful creator.

I used the term postulate carefully. Unlike many, both believers and nonbelievers, I recognize it as a postulate. Still, I consider the probability that the universe has a purposeful creator to be extremely small.

I do not want to call my position atheist, however, because that implies not only certainty but also disrespect for religion and religious teachings. Let us call it *strong agnosticism*, though I mean the term somewhat differently from the science writer Timothy Ferris [1997] of the University of California at Berkeley. To Ferris, strong agnosticism is the belief that the existence of God cannot be proved; weak agnosticism simply states, with Pellegrino, that we cannot know. I go farther than Ferris, however: Strong agnosticism, to me, means strongly doubting the existence of God —indeed, being nearly certain of the nonexistence of God—but admitting the possibility of being wrong. I agree with Ferris, however, that the strongly agnostic position—mine or his—is open to disproof: All God has to do is appear and work some indisputable miracles, that is, miracles that cannot be duplicated and are not open to alternate interpretations.

Do I think that all believers are irrational? No. Only scriptural literalists and others who are certain that they know details about God, God's law, or God's will. I am sympathetic with those who postulate that the universe has a purpose, even if we do not know what it is, but I cannot agree with them because I can find no evidence to support that conclusion.

Determinism

The Marquis de Laplace, a French astronomer and mathematician, believed that, if you could know the mass and state of motion of every particle in the universe, even for an instant, then you could unerringly predict the future exactly. That was a common belief in the days before the development of quantum mechanics and was consistent with Newtonian physics. Quantum mechanics, however, suggests that motion at the subatomic level is random or at least has a random component to it.

In my view, Laplace was right, and the universe is completely deterministic. In the same way, biological systems (like us!) behave according to the laws of chemistry and physics. To me, free will is an illusion brought about by the seemingly large array of choices open to us. In fact, I think, we have no more free will than a chess-playing computer, which seems to choose from a myriad of possible moves but is clearly a deterministic system that has no will of its own. The remainder of this chapter is devoted to supporting this claim, which some will find extraordinary.

Newtonian determinism convinced some observers that we have no free will. If you think of your brain and your body as physical systems and you believe in the kind of determinism that Laplace espoused, then you can come to no other conclusion. Quantum mechanics seemed to many to give back our free will, [Swinburne, 1996] though that argument has always escaped me. Indeed, most of us would like to think that we would invariably make the same moral choices time and again if presented with the same moral problems under the same conditions. If you allow a random component to your thinking, you might find yourself coming up with the "wrong" choice some of the time. Or you might be pricked by a pin and find it pleasureable some of the time. I think that there is no need to worry about that, however. A large ensemble of quantum mechanical particles, such as your brain, evolves with minimal uncertainty, so I assume your thinking will be much the same each time, provided that the conditions are the same.

In this sense, it may not matter much whether the underlying quantum mechanics is deterministic or not. That is, if you have a large enough number of particles, then you can make predictions with great accuracy. For example, consider a lump of 10 billion (10×10^9) radium atoms. The nuclei of radium atoms *decay* by emitting alpha particles. We can never predict when a given radium nucleus will decay, however; all we can do is state that it has a certain probability of decaying each second. We usually express this probability as a half-life: the time that has to elapse before half of a large number of atoms has decayed. After one half-life, approximately 5 billion will have decayed.

We may make this statement with considerable confidence. The uncertainty of the number 5 billion is about 100,000. That is, there is a 95 percent probability that the number of decayed radium atoms is between 5 billion minus 100,000 and 5 billion plus 100,000. There is well over 99.9 percent probability that the number of decayed radium atoms is between 5 billion minus 200,000 and 5 billion plus 200,000. To put it another way, we know the number 5 billion with an uncertainty of about 1 or 2 thousandths of 1 percent. Your brain consists of several billion neurons and vastly more atoms and molecules; unlike the radium atoms, they do not act not independently. It seems very likely therefore that your thinking or your reactions to stimuli evolve with minimal uncertainty.

But does quantum mechanics do away with determinism? I think not. To illustrate why, let me propose a thought experiment. Figure 4 shows a ball rolling around in a bowl. The bowl is placed in a hole in a table, and the lip sticks up above the table top. The highest points that the ball can attain are called the turning points. Let us assume that the turning points are higher than the table top but lower than the top of the bowl. In classical or Newtonian mechanics, the ball can never escape from the bowl because it does not have enough energy to reach the lip. If there were no friction, it would roll back and forth forever, never rising higher than the turning points. Suppose now that the bowl represents the nucleus of a radium atom. The ball is an alpha particle. Provided that the turning points are higher than the table top, the alpha particle has a finite chance to

255

escape from the nucleus; when it does so, we say that the radium nucleus has decayed.

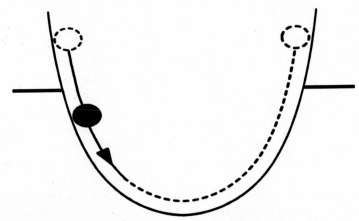

Figure 4. Particle in a bowl. The particle rolls between the turning points indicated by the dashed circles. The turning points are above the top of the table but below the top of the bowl. In classical mechanics, the particle is confined to the bowl, as long as the turning points are lower than the top of the bowl. In quantum mechanics, however, the particle has at least a chance of escaping from the bowl and rolling onto the table top. How does it get out?

Why does a particular radium nucleus decay when it does? We do not know. But I find it very hard to argue that the nucleus decayed when it felt like doing so. Rather, it seems to me, something most likely "told" it to decay when it did. (Sorry for the anthropomorphisms, but that is how physicists talk.) Roughly, something either kicked the particle out or lowered the top of the bowl for an instant, and the particle escaped.

Einstein said that God does not play dice. I think, rather, that God (allegorically speaking!) indeed plays dice, but that the roll of the dice is deterministic. Specifically, you roll the dice around in your hand, but their motion is determined purely by gravity and the forces you impart to the dice with your hand. The way the dice leave your hand depends on the way you put them into your hand and on the way you rolled them around in your hand before you released them. The way they end up on the table depends on their positions and velocities at the instant they left your hand. Chance is not involved, except in the sense that we cannot predict the motion of the dice exactly. In the same way, I doubt that the quantum world is governed by pure chance and argue instead that physics must deep down be purely deterministic. The appearance of chance is simply a statement of our lack of knowledge, our inability to specify all the possible parameters.

The argument that physics is at root deterministic is not an idiosyncratic invention of mine, and it is more than an extrapolation from a simple thought experiment. Indeed, there has always been an undercurrent of determinism and an uneasy feeling that the usual interpretation of quantum mechanics is deficient because it gives no clear idea of the reality of a quantum system. Bohm and Hiley [1995] argue that the usual interpretation prevailed over a realistic alternative partly for "sociological" reasons. Honesty compels me to agree that here postmodernism has a point: At the fringes of science, when there is no evidence to favor either interpretation over the other, the culture has chosen one.

Bohm and Hiley further argue that the usual interpretation gives no reality to a particle until it is actually observed by a conscious observer, so that interpretation precludes the possibility of even talking of an underlying theory. Many think that it is unreasonable to assume that a conscious observer is needed before an interaction can be deemed to have taken place. Bohm, Hiley, and other like-minded physicists have therefore developed theories of quantum mechanics that are at root deterministic, but so far there has been no way to distinguish between these theories and the prevailing interpretation. The arguments are abstruse and in a way peripheral to our main purpose. I will expand on them in an Appendix. If you find the

257

going heavy, then go straight to the next section, "Damasio and Descartes," but bear in mind my contention that even the subatomic world is purely deterministic.

Damasio and Descartes

René Descartes was born about 30 years after Galileo. He lived at a time when the model of the Heavens was a giant clockwork. The science of mechanics was being developed, and that implied a mechanistic view of life as well. Descartes thought of himself as a thinking being and rejected the idea that humans were mindless automatons (I think; therefore I am). In order to support his view, he concluded that the mind or the soul was made of different stuff from the body. He located the mind in the pineal gland, an organ that, in some vertebrates, resembles an eye. In humans, the pineal gland is not light-sensitive but is neurally connected to the eyes. The pineal gland secretes the hormone melatonin and is probably related to the biological clock, that is, our sleeping and waking cycles. Its function is only beginning to be elaborated, but there is no evidence that the soul or mind resides there.

In postulating that the mind or soul differed from the body, Descartes argued that mind-stuff differs from ordinary matter. In particular, it has no mass or dimension and cannot be divided like ordinary matter. The mind and the body interact, but generally the mind controls the body. Sometimes the mind can operate independently of the body. The mind can be affected by passion, but it can also rise above passion and operate rationally and ethically.

* * *

The neurologist Antonio Damasio thinks that Descartes has got it all wrong. Damasio is Professor and Head of the Department of Neurology at the University of Iowa College of Medicine and Adjunct Professor at the Salk Institute for

Biological Research. He is the author of the well known book *Descartes' Error*. [1994] Damasio considers the brain to be an integral part of the body and thinks that bodily reactions influence so-called rational decision making.

Damasio and his colleagues study patients with injuries to specific parts of their brains. They find, for example, that patients with lesions in a certain location in their brains cannot recognize color. [Damasio and Damasio, 1992] More interesting, they cannot even conceive of color, nor can they remember anything in color. Other patients can match colors but not name them. Yet other patients lose the ability to recall whole classes of words, such as those that refer to animals or tools, which they nevertheless clearly recognize and can describe accurately. These patients have no other evident impairments in their ability to communicate.

Evidence such as this moves Damasio to postulate that the brain does not store images but rather consists of "convergence zones" that store different recollections of the same object in different places. Color is stored in one place, shape in another, class of object in another, and so on. We can recall or recognize an object only if the convergence zones are intact and can transmit the sensations to the cerebrum, where the names of the objects are stored. This information is transmitted through certain other structures that link the convergence zones to the cerebrum.

Sensations cause certain patterns in the brain. Recollections cause similar but weaker patterns. These weaker patterns can nevertheless cause sensations. For example, if you think of a frightening experience, you may break into a sweat or your heart might start pounding, precisely imitating your reaction when the real experience occurred. Thus, the brain and the rest of the body are closely connected, sensations flow both ways, and there is no evidence of a disembodied mind dispassionately watching what is going on.

Damasio postulates that bodily responses, because they are so closely connected with patterns in the brain, actually guide the

259

rational mind to come to a decision. In a review of Damasio's book *Descartes' Error*, Michael Gazzaniga [1995] of Dartmouth University points to related research that suggests that patients with very severe spinal cord injuries find it hard to feel powerful emotions. Indeed, I remember once seeing a television program about rehabilitation of quadriplegics who had recently been injured. I was surprised at their lack of anger, which, according to Damasio's hypothesis, could be because their injuries prevent sensory information from progressing from their bodies to their brains.

You can see for yourself what Damasio is talking about. Try to recall something that made you furious. Now sit back, relax, and make sure that your limbs are loose, your brows not knitted, and your fists not clenched. Smile. I find it difficult to feel angry. I can, however, become extremely angry when I let myself frown, grit my teeth, and clench my fists.

In support of his hypothesis, which he calls the *somatic marker hypothesis*, Damasio describes a set of experiments involving *skin response*. Skin response is, roughly, measuring how sweaty you become when you are startled, frightened, or under pressure. It is one component of a lie detector test. The actual measurement consists of measuring changes of the electrical resistance of your skin in response to a stimulus.

Damasio and his colleague Daniel Tranel compared the skin responses of a control group with patients who were known to have frontal lobe damage and are often unable to make rational decisions. When these patients were startled, they had normal skin responses, that is, equal to the responses of the control group. On the other hand, when they were shown what Damasio calls "disturbing" images, the patients who had frontal lobe damage displayed an absolutely flat skin response. Both the controls and other patients with brain damage elsewhere showed marked skin responses. Yet the patients with the frontal lobe damage were able to describe the contents of the disturbing images in detail and use emotional terms like fear or disgust appropriately. They understood the contents of the images but did not internalize them. Indeed, one patient knew how he was "supposed" to feel but noted that he did not feel that way. Damasio thinks that is because, literally, he could no longer feel the images in his gut.

Finally, Damasio describes what he calls the gambling experiments. The subjects of the experiment were asked to play cards. They were given some play money and four decks from which to draw cards, one at a time. They could draw from any deck but were not taught any other rules. Gradually players figured out that, if they played from decks A or B, they would generally win $100 in play money on each draw. If they played from decks C or D, they would generally win $50 on each draw. Occasionally, however, they would be assessed a fine, and the fine from the $100 decks was so large that you ultimately lost money if you played from those decks. Only if you played from decks C and D could you be assured of winning in the long run.

Even though it was impossible to make a precise calculation of the probabilities, players internalized the rules of the game within 30 draws. That is, they apparently figured out that decks A and B were risky, and C and D were less so. Thereafter, the control group generally played from decks C and D (the $50 decks) but occasionally took a chance and played from decks A and B (the risky $100 decks). By contrast, the frontal lobe patients generally played from decks A and B, and ultimately lost their money. If they were allowed to borrow, they continued to play from decks A and B, and continued to lose.

Patients with frontal lobe damage behaved precisely oppositely to the control group. One patient whom Damasio calls Elliott had had a brain tumor and as a result had a portion of his frontal lobe removed. From then on, Elliott could not be counted on to make what we would call rational decisions, even though his faculties seemed otherwise intact. Elliott lost his job, invested badly, went bankrupt. When Elliott played the card game, he went broke. A few months later, he played the game again and went broke again. Elliott understood the game perfectly, played attentively, and wanted to win. Yet his choices were inevitably disastrous; they mirrored his choices in life.

Hanna Damasio, a neurologist with the University of Iowa and the Salk Institute, suggested monitoring the skin responses of both patients and controls while they played the game. After the controls had figured out the rules, they showed a marked skin response whenever they decided to gamble and play from one of the high-risk decks. That is, they learned to fear playing from these decks. The frontal lobe patients, by contrast, showed no skin response whatsoever.

Antonio Damasio postulates that the normal brain uses a nonconscious process to estimate the danger of any proposed action. That estimate induces a somatic (bodily) response, as is evidenced by the skin response. The skin response is not a mere epiphenomenon (a phenomenon that accompanies another but has no other significance). To the contrary, the brain feels the somatic response, and that response guides the brain in preparing for making a conscious, cognitive decision. Patients who cannot feel the somatic response are unable to make rational decisions because they have no fear of making the wrong decision. By rational, I mean that they are unable to act in a way that will give them the highest probability of success as they understand it.

Why do these patients make consistent but wrong decisions instead of random decisions? Damasio suggests that their lack of fear limits their perspective to the immediate future. If you have no fear of delayed consequences, then you will choose immediate reward rather than worry about future losses. The odds are that you will sometimes get your $100, so you draw from decks A and B, even though you know intellectually will suffer long-term consequences.

Whether Damasio's proposed mechanism is right in detail or not, he seems to have shown clearly that somatic responses are necessary for what we usually call rational decision making. Indeed, even trivial decisions may need a somatic response. Damasio describes a frontal lobe patient who spent the better part of half an hour unable to choose between two times for an appointment; the patient never decided but finally accepted the suggestion of the physician.

Damasio has made two discoveries that concern us here. First, he and his colleagues have identified specific locations for specific functions in the brain. Second, they have shown that a somatic response is necessary for rational decision making. That is, they have shown that the brain is more wholly integrated with the body than had been previously thought. The brain cannot think rationally without information from the physical sensations of the body; those brains that are forced by injury to do so almost invariably make irrational decisions. In short, there is no evidence for a nonmaterial observer in our brains, and there is nothing mystical going on inside our brains. Thus, I suggest,

> *The brain is an organ like any other, and it does not*
> *rely on any nonphysical entities for its operation.*

The brain is merely more complicated and less well understood than most other organs.

Frank Vertosick, Jr., [1996] is a neurosurgeon at Western Pennsylvania Hospital in Pittsburgh and majored in theoretical physics as an undergraduate. He tells of a teen-aged patient who had accidentally been shot in the head with a pistol. The patient's frontal lobes were severely damaged. Before he was shot, the patient had been a difficult child and had a long arrest record. Vertosick operated and cleaned out the damaged portion of the patient's brain. The patient recovered and, in contrast to his behavior before the shooting, was doing relatively well in school and no longer argued or displayed sudden anger. Vertosick also mentions the case of the railroad worker Phineas Gage, who was injured when an accidental explosion shot an iron bar cleanly through his head and damaged his frontal lobes. [Damasio, 1995] Gage, who had been responsible and well-liked before the accident, recovered physically but became obnoxious and irresponsible afterward. The teen-aged patient changed in precisely the opposite way from Gage, but both

263

changed radically as a result of damage to their frontal lobes. Vertosick says flatly that our sense of right and wrong, our moral restraint, and our holding to the conventions of society derive from our brain no less than do our physical actions.

* * *

A digression on reductionism: The mathematical physicist Steven Weinberg [1994] entitles a chapter, "Two cheers for reductionism." Accused of being an uncompromising reductionist, he defends himself by claiming to be a compromising reductionist. What is all the fuss about?

Reductionism is often used to imply oversimplification. For example, if you argue (as Dawkins has been accused of doing) that all human behavior is biologically determined, you will be accused of reductionism. But reductionism does not have to mean oversimplification. Rather, as Weinberg puts it, it means the conviction that scientific principles are underlain by deeper principles. That is, that molecular biology is the foundation for environmental or population biology, chemistry is the foundation for molecular biology, and physics is the foundation for chemistry.

Reductionism does not mean that the most fundamental problems are necessarily the most interesting or even the most important. Neither does a reductionist approach to science necessarily have practical significance; the electrical engineer should not abandon her circuit equations in favor of the quantum-mechanical wave equation, for example. To do so would be impractical, even though it is clear that the circuit equations are in principle derivable from the wave equation.

A reductionist would argue, with Peacocke, [1993] that matter may be organized naturally in a series of levels: atom, molecule, organelle (a functional subunit, or "organ," within a cell), cell, organ, organism, ecosystem, and so on. These levels are not completely isolated from one another like a set of Russian dolls, but they interact weakly enough that any one level

264

can often be described fairly accurately without reference to other levels. The result of such a hierarchy of levels can be *emergent properties*, or properties that emerge when we create a complex system from simpler entities. For example, one or ten or perhaps 100 water molecules are not wet or viscous, but the water in a glass is wet and viscous. Wetness or viscosity is an emergent property, which appears only when we create a system of many water molecules. We could probably not predict wetness or viscosity from the properties of hydrogen and oxygen, the atomic constituents of water, nor even from the properties of a single water molecule; in this sense wetness or viscosity is an emergent property. Nevertheless, we can explain, or *retrodict*, wetness or viscosity on the basis of what is known about individual water molecules. Properties that we call emergent are little more than properties that we can observe but cannot predict, though we can often describe them in terms of underlying principles.

I disagree with Peacocke's characterization of reductionism as *nothing but-ism*, however. I do not think that only atoms or subatomic particles are real, whereas organisms are somehow not real because they are composed of atoms—that is, I do not think that organisms are *nothing but* collections of atoms. Each level of reality may be characterized and understood on its own, separately from the others, though there are interactions between levels. Indeed, what happens at one level can influence what happens at a lower level, as when the struggle for survival among organisms selects the genes of the successor organisms. In short, biology is not a special case of physics. Nevertheless, biology can be understood in terms of physics, and I see no need to invoke any mechanism, such as God or a *life force*, that makes biological systems anything more than complex physical systems.

Specifically, reductionism means that almost anything can ultimately be explained in terms of a deeper or more nearly fundamental theory. Science would not progress without reductionism, for without it we would never try to explain anything in terms of anything else. Newton was being reductionist when he reduced Kepler's three laws to a single universal law of gravitation and also when he deduced that what holds us to the earth was the same force that holds the planets in their orbits.

Weinberg's title has a certain flair to it, but a more accurate title might have been, "Three cheers for reductionism, properly defined."

I have been careful to distinguish between reductionism and oversimplification because I am a compromising but probably incorrigible reductionist. I suspect that most basic scientists are reductionists, because, as I have noted, reductionism is what we practice every time we try to explain any observed phenomenon in terms of something simpler. As I have suggested, we have so far succeeded in understanding biological processes in terms of more-basic chemical processes, and we understand chemistry in terms of more-basic atomic physics. We can explain how the kidney works in terms of a physical and chemical process called osmosis. We understand the quantum mechanical interaction that stimulates the retina of the eye, and we understand how that stimulation is sent to the brain. We understand much about how the muscles work, how digestion takes place, how the immune system works. We are beginning to understand how the brain works, on the molecular level, on the cellular level, and on the integrated or systems level. I will not be surprised if we can someday understand music and art in terms of psychophysics.

*   *   *

I have argued above that, at the deepest level, all of physics is probably deterministic. I do not mean that there is (or could be) a being who has foreknowledge of everything that will happen, nor that we can determine the future in any realistic way. Many authors seem to me to confuse determinism with predictability. Something that is deterministic may not be predictable because, for example, we do not know the values of

266

the stray gravitational or magnetic fields caused by distant bodies. We are forced to treat these fields statistically (if we need to worry about them at all), but that is only because we cannot measure them.

By deterministic, then, I do not mean predictable in practice. Rather, I mean that nothing happens by accident, that every effect has a proximate cause. I think that this applies to biological processes, since they are nothing more than complex physical and chemical processes. Even if I am wrong and quantum mechanics is not at root deterministic, the result of a large number of quantum mechanical events is nevertheless predictable within very narrow margins, and, therefore, biological systems must behave highly predictably in principle even if not completely deterministically. In short, I think that everything that goes on inside our organs is a physical process, and that all physical processes are deterministic or highly predictable. Since our brains are organs, I cannot avoid the hypothesis that

> *Everything that goes on inside our brains is determined by the laws of the universe and by nothing else.*

That is, we are no more in control of our brains than we are of our livers. What happens inside our brains is solely the result of physical and chemical processes. Just as temperature is an emergent property that results when billions of molecules are in motion, what we call consciousness is probably an emergent property that results when billions of neurons are interconnected in a single system. Indeed, the philosopher Daniel Dennett [1996] comes close to arguing that consciousness is the self-awareness of a brain that can form concepts by means of language.

The manner in which the brain operates is a function both of the physiology of the brain as such and of our previous experiences. In this sense, both nature and nurture are important, but I am now convinced, largely by the studies of identical twins, [Holden, 1980; Wright, 1995] that nature is a major influence.

I have come to this conclusion only with difficulty. In the 1960's, my friends and I believed firmly in the power of environment to shape our behavior. At the time, I could never have entertained the idea, for example, that Blacks could be inherently better basketball players than Whites, even though it was perfectly clear that various African ethnic groups have different morphologies from Europeans, who have different morphologies from Eskimos, for example. Why, we used to ask, was the National Basketball Association dominated by Black players? Because Blacks lacked other opportunities or were taught better for some cultural reason. Why did little boys generally prefer tools and little girls prefer dolls? Because those preferences were taught. It was not politically correct to suggest that there could be any other explanation. We regarded as racism or sexism any suggestion that there could be significant inherent differences among races or ethnic groups or between the sexes. In short, I used to reject certain theories because I did not like their consequences (see "Taste," Chapter 2). (I am not arguing that Blacks are better basketball players than Whites: only that the theoretical possibility exists. It would not take much of a difference on average to ensure that the best few hundred basketball players in the country were mostly Black. If you think this is racism, ask yourself who would more probably win a 500 kilometer dog sled race—the average Eskimo or the average Tutsi. Eskimos are generally short and compact, for conserving heat, whereas Tutsis are generally tall and lean, for dissipating heat. My bet is therefore on the Eskimo, assuming that both are equally trained. Similarly, I would bet on the Tutsi in a one-on-one basketball game.)

As the evidence has poured in, I have had to get used to those theories, whether I liked them or not. Even A. S. Neill, the founder of the Summerhill School in England and a very doctrinaire reformer, has changed his mind on this point. He used to think that boys tinkered with bicycles and girls sewed or knitted because of "custom and training." At Summerhill, Neill and the staff tried to minimize sex differences and allow the children to develop as they pleased, which Neill called allowing them to grow up in freedom. After years of observing children at his school, however, Neill concluded that "freedom does not alter the innate predilections of the sexes." [Neill, 1966, pp. 66-67] Neill's evidence is anecdotal and based on the assumption that the children were really raised in "freedom" and not especially tainted by the world outside Summerhill. I do not know whether he is right, but I can no longer reject the possibility.

The evidence cited by Holden, Wright, and others shows the importance of biology to our behavior. Other evidence suggests that biology is purely a complex chemical and physical process. It is but a short step to the conclusion that we are solely biological beings and that all our behavior is determined. We think that we have free will only because of the large number of possibilities open to us, but the possibilities we actually choose are at root determined by the underlying processes of chemistry and physics.

* * *

My friend Kim argues that I have reduced human behavior to biology, chemistry, and physics, but I have not shown that there is not something else—something additional, possibly external to us, that gives us free will. True: I have not, and I cannot. Probably we will never be able to show anything of the kind with certainty. My argument, however, is that we can reduce an enormous amount of human behavior to biology, and no one has shown any evidence that there *is* something else. Since we can explain behavior without that something else, it is a hypothesis we have no need of. In short, I deal only in observables, and Kim's "something else" is not observable.

Kim insists further that he has ample evidence for his own free will, but that evidence is only that he feels as if he has free

269

will. That is flimsy evidence for a scientist; it is no better than the mystics' arguments or Ricky's maxim, "If I believe it, then it is true." I think it more likely that Kim has the illusion of free will because he often has a great many choices available to him. The choices he makes, however, are as constrained by his biology as were those of Damasio's patient Elliott. Elliott made choices, but they were apparently the choices he had to make, not the choices he wanted to make.

In short, if you agree, as Kim does, that we are biological systems and that the underlying physics is deterministic, then I think you have to admit that I have made, at least, a *prima facie* case against free will. The burden of proof that there is something else therefore shifts to Kim, and he has yet to make an effective rebuttal.

My friend Davy similarly does not like this section and asks whether I think therefore that we must not hold people responsible for their choices. He knows that he should not judge a scientific theory on the basis of his feelings about it, but he asks whether, to be consistent, I would abandon the judicial system. My answer, which I discuss in Chapter 9, is, "No." We still have to behave as if we had free will, and we have to control unruly elements in our society, whether or not they have free will.

If something is in me which can be called religious
then it is the unbounded admiration for the structure
of the world so far as our science can reveal it.

ALBERT EINSTEIN

Chapter 8
The magnificent structure of nature

We have come a long way. I have tried to convince you that,
contrary to postmodernist assertion, there is objective reality or,
if you prefer, objective truth that exists independently of the
observer and the belief system of the observer. I have argued
further that the only way to get at that truth—more precisely the
only way to approximate it—is through empirical observation.
That observation must not be casual, however; empiricism must
be supplemented with reason and care, or else you fall into twin
traps of believing what is agreeable to you and of relying on
selectively chosen anecdotes or vague and unprovable
hypotheses as supporting evidence.

The hypotheses of religion, and (as I have tried to show)
they *are* hypotheses, must be treated the same way as any other
hypotheses: They must be examined critically and tested. That
is, we must ask—we have an intellectual obligation to ask—are
the hypotheses supported by the available evidence?   In this
book, I have tried to show that they are not.

I have dismissed what I called "popular" beliefs such as the
belief in signs or miracles on several grounds.   First, most
presumed miracles can be explained or accounted for without
assuming divine intervention. Storms and other natural disasters
are just those: natural disasters and not acts of God. I firmly
reject the arguments of those who give God credit for all that is
good and ignore all that is bad; they are using evidence
selectively in order to bolster a belief that they must intend to
hold onto come hell or high water.

Similarly, I cannot accept the kind of wishful thinking that there must be a God because, otherwise, there would be no purpose to our existence, no fixed values, no universal code of morality. You cannot arbitrarily hypothesize, for example, a universal code of morality and then use the presumed existence of that code to "prove" that there must be a God. This hypothesis is not obviously true and requires evidence to support it. Basing one unsupported hypothesis on another, equally unsupported hypothesis is not progress.

Even though they are tantalizing because they appear statistically significant in a way that anecdotes do not, I cannot accept the equidistant letter sequences in the Book of Genesis as a sign from God, particularly in light of the strong circumstantial case that the Bible was compiled from a multiplicity of sources, which are often at odds with one another. In addition, it now appears that Witztum and his colleagues may well have adjusted the data until they achieved the result they wanted.

In the Western world, a great many people nevertheless think that the Bible is the literal word of God. The myriad of errors and inconsistencies in the Hebrew Bible and in the Gospels ought to deliver a death blow to that belief: At most, the Bible is the word of God as interpreted and distorted by generations of oral tradition and then by later redaction. The Book of Jonah is so obviously a fiction that I am astonished any time I hear someone argue for its literal truth. The Gospels are not contemporaneous accounts of the life of Jesus, and they are unsupported by external evidence. Each successive account may be no more than an embellishment of the preceding account; only the first account is even roughly accurate, and there is no independent evidence for supernaturalism. As important as the Bible is, it is not the literal word of God.

The major problem for those who believe in a benevolent God is the existence of evil and misfortune. A rough accounting suggests that there are at least as much evil and misfortune in the world as there are benevolence and good fortune; religion itself has been responsible for much evil and continues to be to this day. The Bible gives no answer to this problem. Specifically, the most commonly cited theodicy, the Book of Job, offers little or no help. The comforters mostly blame the victim, which is not at all enlightening. God himself never once claims to be just: only powerful. He seems to be saying, "Might makes right," a sentiment that our society has long abandoned as moral justification.

We can, however, find a potential source of evil in biology. When we see analogies to evil in the animal kingdom, we are properly reluctant to classify them as evil. In our minds, only humans can perpetrate evil. I conclude, therefore, that evil does not exist except insofar as we define it. It needs little or no explaining unless we hypothesize a benevolent God. That is, the God hypothesis hinders our understanding of evil, rather than helps it.

I have therefore found wanting almost all the arguments of laypersons and the clergy alike. How well do philosophers fare? Not well: Their proof texts are not as old as those of the scriptural literalists, but they seem as dated, and, except for a few philosophically minded scientists, philosophers of religion seem as unwilling to incorporate the discoveries of modern science into their worldviews as are the Biblical literalists.

The Ontological Argument makes no sense to me. It is based on the unsound premise that any valid logical argument must necessarily apply to the physical world. It is a wild extrapolation, and it is not testable.

The Argument from First Cause fares slightly better, but only because of the empirically supported claim that the universe has a finite age. If it has a finite age, then it probably had an ultimate cause. There is, however, no evidence that the ultimate cause was purposeful, so the Argument from First Cause ultimately fails as well.

The Argument from Contingency fails for much the same reason that the Ontological Argument fails: It defines a necessary being by analogy with the logical concept of a necessary proposition. That there are logically necessary

273

propositions has no bearing whatsoever on whether or not there are physically necessary entities, so the argument is wholly unsound.

The Argument from Design assumes without evidence that the universe was designed for a purpose. It is largely circular in this sense. As a general argument, it is weak. The Argument from Evolution, by contrast, is firmly grounded in the fact that the right edge of the graph of complexity moves almost inexorably farther to the right as (geological) time progresses. The existence of periodic mass extinctions, however, argues strongly against the claim that the universe was created with intelligent beings (not necessarily us) as its ultimate goal. The Anthropic Principle, which follows from the Argument from Evolution, seems to me to be completely circular, and I cannot take it seriously.

The Argument from Mathematical Physics depends on whether you think there is order at the deepest levels of reality. I do not see any reason to ascribe the existence of order to a purposeful creator, and I think that the Argument from Mathematical Physics fails for much the same reason that the Argument from Contingency fails: There is no reason to believe that the universe was created by a mathematician just because we can describe it by mathematics.

It was frankly a surprise to me that philosophers take the Argument from Religious Experience seriously. I see no difference between that argument and Michael Crichton's argument that a miracle happened to him when he met his future wife. There is not one shred of evidence, credible or otherwise, that mystical or religious experiences are objectively real and not hallucinations or other well understood mental phenomena. That is, although the mystical experiences seem real, no one has ever devised an argument that could be used to distinguish them from well known and well understood artifacts such as hallucinations and dreams.

I conclude that the evidence in favor of a purposeful creator, let alone a benevolent God, is so weak as to be virtually

274

nonexistent. Indeed, it is so weak that we are justified in arguing that the hypothesis has been falsified. There is almost certainly no purposeful creator and certainly no benevolent God.

What then do I believe in?

Believe is a strong word. I do not *think* that the universe had a purposeful creator. I am *almost certain* that God does not intervene in our affairs, that there is no absolute code of morality, and so on. I probably believe these things as firmly as all but the most rigid literalists believe the very opposite. I differ from the literalists, however, in my admission that I could be wrong and in my continuing search for the evidence, either way. In short,

> *I try to believe what I have to believe, not what I want to believe.*

I am nearly convinced, partly by the thought experiment of the particle in the bowl, that the universe is completely deterministic. Even if it is not, the wavefunction of a complicated quantum system such as a brain evolves with almost perfect predictability. I suspect that far more of our personalities are determined by the physiology of our brains than is generally recognized. Indeed, my statement that the universe is deterministic compels me to hypothesize that all our actions and thoughts are determined once and for all by the laws of nature. In this sense we have no free will: Free will is an approximation that we make because we can do nothing else; it is a concept that we developed because we seem to be free and have a great many choices open to us. But I doubt that we are free in the strictest sense of the term.

Some people find this argument very threatening: It might imply that mind is an epiphenomenon, that is, the result of physiological processes in our brains and bodies, and nothing more. That there is no purpose to our existence. That one day there will be no more humans, no earth, no universe as we know it. To me, however, these are plain physical facts with no moral

275

or ethical content. The fact that we do not have immortal souls does not justify unethical behavior. I might like the world to be otherwise, but it is not.

What then can I propose to replace theism? First, the knowledge that the universe is intelligible. As a scientist, I see or read about phenomena that must seem like miracles to laypersons and certainly seemed like miracles to the ancients. The ancients postulated a god or gods to explain the natural order. Today, however, we find the universe understandable in terms of physical laws and have no need to invoke supernatural powers. In place of theism, I propose what Einstein called a *cosmic religious feeling*, an "unbounded admiration for the structure of the world so far as our science can reveal it." To Einstein, the awe and humility he felt in the presence of the "magnificent structure" of Nature were a genuinely religious feeling, but it was firmly grounded in reality and required no anthropomorphic God.

Second, without a literal belief in a god who dictates moral codes or guides us along our paths through the universe, I propose the idea that we are grownups, on our own and responsible for ourselves, not children for whom someone else is responsible.

Finally, I propose, to those who want it, a religious humanism that is human centered, not God centered. In this view, our lives have meaning, but it is meaning that we and our communities give them, not meaning that is derived from a supernatural source. We have to act as if we had free will, because we can do nothing else. But we and our communities have to develop our own ethics. There are no moral imperatives and no universal code of morality, no automatic rewards for good deeds, no automatic punishment for bad deeds, no God looking over our shoulders. All we can do is strive to improve ourselves and our world, and we are completely on our own. Far from despairing, however, I consider hopeful the facts that medicine and sanitation have improved our health and longevity; science and technology have given us shorter working weeks, more abundant food and resources, and more leisure; and our political systems have given us more freedom and dignity. The power to improve the system further and to extend our good fortune to the rest of the world is in us and our own rational thinking, not in God. To put it in theological terms, we must seek our salvation in this world, because there is no other.

In the next chapter, I will conclude with a series of questions and answers that I hope will clarify some of these points better than a didactic presentation. The questions are real in the sense that they are questions people have asked me over the years. Many of the answers and the follow-up questions are paraphrases or composites of real conversations as I remember them. Some go back to college; some are relatively recent. Other answers are padded slightly with later material or material I acquired as a result of this research.

A man said to the universe:
"Sir, I exist!"
"However," replied the universe,
"The fact has not created in me
A sense of obligation."

STEPHEN CRANE

Chapter 9
Questions theists ask

You do not believe in God at all?
Not literally, no.

Then there is nothing higher than humans?
I do not know what you mean by "higher," but I know of
nothing more intelligent than humans.

*Isn't it sort of arrogant to think that we are the highest form
of life?*
That is not what I think. I simply do not know of the
existence of any entities that are more intelligent than we. By
intelligent, I mean able to solve certain kinds of problems.
Unfortunately, we do not always use that intelligence wisely, but
that is another issue.

As to your use of the word "arrogant," I consider it instead
somewhat humbling to realize that we are the result of a series of
accidents: the very opposite of arrogant. Indeed, I consider it
just a wee bit arrogant to believe that the universe was created
with us in mind. We live on one of several planets of an
ordinary star in no particularly central place in an ordinary
galaxy among billions of galaxies. Frankly, I think it strains
credulity to argue that the universe was created with some
purpose, of which we are an integral part.

Doesn't your scientific approach to everything rather dehumanize us?

To the contrary, it recognizes us as what we are: a species of intelligent animal. Jacob Bronowski [1973, p. 374] calls your assumption, that science is dehumanizing, "tragically false." His book is the transcript of a television series, and he was speaking from the site of the concentration camp and crematorium at Auschwitz. Bronowski notes that 4 million people were cremated there. "And that [the Nazi atrocities] was not done by gas," says Bronowski.

> It was done by arrogance. It was done by dogma. It was done by ignorance. When people believe that they have absolute knowledge, with no test in reality, this is how they behave.

Science is amoral. Its discoveries can be used for good or for evil. In that respect, it is no different from religion.

You talk about truth; what about beauty?

Beauty is an internal reality. If we think something is beautiful, then it is. I mean that independently of whether beauty is in the eye of the beholder or whether some concepts of beauty, such as symmetry, harmony, or matching colors, are hard-wired into our brains. Even if they are, there is no objective criterion for measuring beauty as there is for measuring mass; different people will get different results if they try to quantify beauty on a scale from 1 to 10, for example. In this sense, beauty is an internal reality, not an external reality.

Walt Whitman wrote a famous poem, "When I heard the learn'd astronomer." Here is the poem:

> When I heard the learn'd astronomer,
> When the proofs, the figures were ranged in columns before me,

When I was shown the charts and diagrams, to add,
    divide, and measure them,
When I sitting heard the astronomer where he
    lectured with much applause in the lecture-room,
How soon unaccountable I became tired and sick,
Till rising and gliding out I wander'd off by myself,
In the mystical moist night-air, and from time to time,
Look'd up in perfect silence at the stars.

The clear implication is that the astronomer and his audience cannot appreciate the beauty of the stars. By analyzing them, the astronomer has spoiled Whitman's experience viewing the beauty of the night sky; by inference, science has taken the beauty out of the world.

I cannot disagree more with this sentiment. As a scientist, I too can look up in perfect silence at the stars. That I have some idea how they work makes them no less beautiful. To the contrary, perhaps I appreciate the sight more than Whitman because I appreciate both their beauty and the beauty of the mechanisms that make them work. Schumann's piano quintet Opus 44 is no less a wonderful piece of music because I know that the body of the violin has to resonate in a certain way with one of the strings or because I know why one note harmonizes with another. The rainbow is no less wondrous because I know how it is formed. Flowers are no less beautiful because I know what the parts are for. Vincent van Gogh's painting Starry Night is no less excellent because I know the difference between subtractive color and additive color.

Far from dehumanizing, science allows us to appreciate beauty on yet another level, a level that is not open to the nonscientist. Einstein [1954] considered being in awe of the sublimity and order in the universe to be akin to a religious belief, and he called it a cosmic religious feeling. Einstein's religious feeling, therefore, was no less than that of any conventionally religious figure. Indeed, perhaps it was stronger because it was coupled with verifiable scientific beliefs.

281

* * *

A very sizeable majority of Americans believes in God. Why do you insist that they are wrong and that you are right?

The available evidence does not support the hypothesis. To the contrary, if the hypothesis is that God is benevolent, the evidence clearly argues against the hypothesis. A benevolent God would never have allowed the Holocaust, unless he was also impotent.

Does God have to be benevolent?

Not at all. In fact, if you postulate God but do not make him benevolent, your case is harder to disprove. It seems to me that assuming God's benevolence is the weakest point in the theist's argument.

I believe that God withdrew himself from us in order to give us free will. He was unable to intervene because intervention would have interfered with our exercise of our free will.

I cannot accept this argument. It is a patch—an *ad hoc* hypothesis—that makes the hypothesis of a benevolent God inherently unfalsifiable. It is designed specifically to let God off the hook. In your view, God is benevolent but lets us torment each other out of principle.

Where does faith fit into your worldview?

There is no place for faith in my universe. At least not faith in the way that you mean it. Rather than have faith in something that may not be true, I think we should question everything, try to figure out for ourselves what is true and what is not. Faith without evidence is a blind alley. It is a refuge for people who refuse to think for themselves.

Bertolt Brecht's play *Galileo* contains a scene where two prelates are invited to look through Galileo's telescope to see for themselves the moons of Jupiter and the stars in the Milky Way. Instead of looking, they apply the dogmas of the Church and deduce that Jupiter cannot have moons: If there were moons orbiting Jupiter, they would have to crash through the crystal sphere that keeps Jupiter suspended. Therefore, there is no point in looking into the telescope, because there is nothing to see anyway. The story may not be historically accurate, but it illustrates how faith can hamstring knowledge or understanding. The medieval Church's insistence on the primacy of faith over reason set science back for generations. Descartes was the most important scientist of the generation that followed Galileo; after Galileo's recantation and his imprisonment for espousing the Copernican theory, Descartes stopped publishing in France and took refuge in Holland.

Forgive me for putting it this way, but I do not believe in astrology, in faith healing or psychic healing, in foot reflexology or iridology, in homeopathic medicine or ayurvedic medicine, in telepathy or clairvoyance or ESP, in past lives or racial memory—nor in a literal God. And all for the same reason: Convincing evidence in favor of any of these propositions has not been found.

But you go to synagogue and observe the holidays.
Yes, and I say in Hebrew, "Blessed are you, Lord our God, King of the Universe, who has brought forth bread from the earth."

Isn't that inconsistent?
Are you being too polite to say "hypocritical"?

OK: hypocritical.
No, because when I say, "Blessed are you, Lord our God," I mean it allegorically. I am expressing in poetry my delight in being alive and in belonging to a group.

Belonging to a group?
Yes. Let's be honest and admit that religious affiliation is one manifestation of tribalism. In my case, I was raised by

Jewish parents. They were largely anti-religious, but I knew I was Jewish and felt a part of that community. In addition, the Jewish-American subculture has influenced my thinking, both through the people I associate with and the books and magazines I read. So when I fast on Yom Kippur and refrain from eating bread on Passover, I do so at least in part to express my connection with that subculture. My gratitude, if you will, to that subculture for providing an intellectually stimulating environment that allows me to read, study, and work things out for myself. Frankly, that is not an option for a great many who belong to other hyphenated-American subcultures, whether Jewish or Christian, such as those sects that adhere to a literal interpretation of the Bible.

Also, I find certain Jewish teachings and values congenial. Take the Talmudic verse that you cannot use the Day of Atonement to atone for a wrong you have committed against someone else; rather, you have to go and make it right with that person before you can atone before God. In one of the synagogues I attend, we recite this verse on the Day of Atonement. It is a far cry from the view, which I cannot accept, that Adolf Eichmann (the man who ran the trains to the Nazi death camps) could have gone to Heaven if he had truly repented before his death. In the Jewish view, actions speak louder than prayers or dogmas.

The influential rabbi Mordecai Kaplan borrowed the word *sanctum* (literally a sacred place or a holy place) to mean the entire constellation of things that are holy to a group. Thus, *sancta* are the sacred texts, heroes, places, events, and rituals of a group. Independence Day is an American sanctum, as is the myth of George Washington and the cherry tree. I observe the Jewish holidays, light candles on the Sabbath, study the Torah, and go to synagogue and recite certain prayers because they are Jewish sancta and not because of any traditional religious belief.

When I visited the Western Wall, one of the retaining walls from the Temple complex in Jerusalem, I knew exactly what Kaplan meant. I felt as if I were in a sacred place. I thought of

284

perhaps 80 generations of my ancestors who longed to travel to Jerusalem, and I was here by virtue of modern technology. I thought of all the blood spilled over Jerusalem, of the martyrs, and I, in my turn, longed for peace.

Isn't that a cop-out? Why not satisfy your belonging to the group without the religious overtones?

Because observing certain religious holidays is one of the things that my group does. If I want to belong to the group, then I have to accept at least some of their practice. I find praying acceptable because I *transvalue* the meanings of the prayers. Reform Jews also change some of the prayers to make them more acceptable to modern people. For example, we no longer say that God gives life to the dead; instead, we say that God gives life to everything. I accept that allegorically. During the Passover seder [the ritual meal], we thank God for delivering us personally from the hand of the Egyptians. I do not think that God literally delivered me from slavery (nor that I would be a slave today if he had not); rather, I look at the seder in part as an affirmation of our belief in freedom. Observing the seder with my relatives and best friends is at least as meaningful to me as going to the fireworks on Independence Day.

In short, when I light the Shabbat [Sabbath] candles, I do it because I want to: because it connects me to my people, my culture, and my history. If I thought I had to light candles because it was the law and for no other reason, I probably would not want to light them, for then lighting candles would be just a sterile ritual.

When you pray, who listens?

I listen. When I pray, I am giving myself a pep talk, like the athlete who says to herself, "Okay! Get the ball back!" On the High Holy Days (the Jewish New Year through the Day of Atonement), we say that those who are destined to live another year are written in the Book of Life, but the Book of Life is not sealed until the Day of Atonement. If you are not written in the

Book of Life, it is perhaps because of some failing on your part, because you can change your situation by acts of repentance, prayer, and charity (*t'shuvah*, *t'fillah*, and *ts'dakah*) before the book is sealed. Now I don't know anyone who believes that literally, but it nevertheless gives us time to pause and reflect on our accomplishments or our attitudes of the previous year and to correct them if possible.

So God is an allegory?

Yes, and I think a lot of liberal clergypersons see him the same way. For example, Ira Eisenstein [1964] is a rabbi and for many years was a leader of one of the branches of Judaism. In *What We Mean by Religion*, Eisenstein notes that each human being has within himself or herself the power to help remake the world. [pp. 68-69] He calls that power the divine in us, or the spark of God in us. Whenever we wish for peace for the whole world or perform an act of kindness, we are using the force that Eisenstein calls God. What is a sin? Failing to be at our best.

Eisenstein says, in italics, *"The Best in us is the God in us."* But I frankly think the sentence should have been turned around: *The God in us is the best in us.* In Eisenstein's view, God is clearly no more than an allegory for our motivations whenever we are being kind, just, or selfless. Religion or religious thinkers often help us to see the way, but the intentions are within us, not external to us.

For instance?

Precisely two days after I sat down to write this book, my wife called from Paris to say, "I am alive." Someone had detonated a bomb in a Metro station she had passed through only twenty minutes earlier. Indeed, she had been on the previous train of the very line in which the bomb had been planted. Should I have regarded it as a warning? Stop work? That is, after all, no more arrogant than the belief that a specific person was deliberately put next to the Oklahoma City blast and left

286

uninjured specifically so that he could help (see "Miracles," Chapter 3).

My wife told me that my son and daughter-in-law were also in Paris that day, though I had thought them to be in Brittany, and she did not know where to reach them. What raced through my mind was a well known Talmudic prohibition: If you see a house burning down the block, you are forbidden to pray, "May it not be my house." Why? First, because a house is already burning and nothing can change which house it is. But also because, by wishing it is not your house, you must necessarily be wishing evil on your neighbor.

Now, it is of course impossible not to wish that my relatives were not involved in the explosion. But the prohibition made me keenly aware that if not my relatives, then someone else's: that the victims were not faceless Frenchmen but real people, someone's relatives. If only the bombers had thought the same way.

Does it do any harm to believe in God?

A rational belief in God does no harm and it can indeed do some good. The American philosopher William James, for example, argued that a religious belief, even if untrue, might have survival value. [Pojman, 1987; Tymoczko, 1996] James gives the analogy of a mountain climber who is trapped somewhere and has to jump to safety. If the mountain climber has faith—even an unreasonable faith—that he can succeed, James believes (without evidence) that he is more likely to succeed than if he is pessimistic about his chances. Further, Tymoczko cites recent research that suggests that people who are mildly overoptimistic are mentally healthier than others. Depressed people may make more accurate assessments of their chances, but such people are not generally regarded as healthy. The reasoning may be circular, or the outcome may be an artifact of the definition of healthy, but the research suggests that religious people may be mentally healthier than nonreligious people because the religious are more apt to be optimistic.

287

Similarly, the newspapers report (as if it were news!) that religious believers are less afraid of death than nonbelievers. Thus, a rational belief in God can provide comfort to the believer, even if the underlying belief is false.

But the irrational (not nonrational; see "Rational, irrational, and nonrational," Chapter 2) belief that your view of God is the only true view can be downright dangerous. A colleague of mine once told me that he was a bigot—his word, not mine—because he believed that those who did not believe in Jesus were absolutely wrong and would not go to Heaven. He hedged a little about those who had never heard of Jesus, but his meaning was clear. I do not agree that he is a bigot, any more than we are bigots if we are scornful of those who believe that the earth is flat, because "bigot" implies intolerance and perhaps active discrimination. But I would agree with him that it is only a short step from an obstinate adherence to your own beliefs to true bigotry. Recently I heard a comment that was very apt: I wish people would spend less time worrying what was said 2000 or 5000 years ago and worry more about getting along today.

* * *

I am Jewish. I had my bar mitzvah when I was thirteen, and I almost never again set foot in a synagogue. Until I read the draft of your book, I thought that you had to believe literally in God if you were to be Jewish—that is, to practice Judaism. I now understand that that is not so: You can go to synagogue and take part in the sancta and be a member of the culture without hypocrisy.

Precisely. And I assume that the same may be said about the more liberal branches of other religions as well. There is no reason that modern rationalists cannot take advantage of meaningful aspects of their religions while discarding the unnecessary baggage of supernaturalism.

* * *

Why do you think that most people believe in God?

It is presumptuous to talk about the motivations of other people. Probably people believe in God for a variety of reasons. Certainly some people invent God in order to give meaning to existence or to explain things for which we have no answers. Some motivations for believing in God are the flip side of Prager and Telushkin's [1986] glib analysis of atheism See "Wishful thinking," Chapter 3). It also seems possible that group selection (see "The biological origin of evil," Chapter 5) or some other Darwinian process has created a predisposition toward a belief in God, perhaps because those societies that were more religious were also more cohesive and therefore competed better against other societies. If you tell a man to sacrifice his life by diving under an elephant and killing it with a spear, he may be more likely to do so if he believes, for example, in eventual reward in the afterlife than if he does not.

Further, as Steve Allen [1993] has noted, religious practices may be, in part, manifestations of mob psychology. People in groups behave differently than individuals alone. As Allen notes, a young woman was unlikely to swoon if she encountered Elvis Presley unobserved. But many young women in Presley's audiences were overcome with emotion. To this day, people deny Presley's death, and "sightings" of Elvis are relatively frequent. Allen considers it likely that religious ceremonies originated in mass meetings for war, harvests, or funerals. Peer pressure did much of the rest.

Indeed, Richard Dawkins [Slack, 1997] has suggested that we may be programmed by natural selection to believe whatever authority tells us, because that was one mechanism for keeping children from, as Dawkins puts it, picking up snakes or eating red berries. That is, for one reason or another, we may have a genetic tendency toward credulity that sometimes overpowers our reason and skepticism. This tendency presumably persists into adulthood, when we become the authority for children but

289

still require an authority figure for ourselves. Even as sophisticated and scientifically educated a philosopher as Ian Barbour, [1990] whom we met in connection with Antony Flew's challenge, discusses different models of God as if God's existence were a settled matter rather than an open question (see "Is religion falsifiable?" Chapter 2).

Still, I cannot help thinking that the belief in God is driven largely by the promise of immortality or, if you prefer, by the fear of death.

The poet Gerard Manley Hopkins converted to Roman Catholicism when he was 22 years old and became a priest 11 years later. He destroyed his early poetry upon his conversion, but he kept writing afterward. His poems are mostly religious poetry, broadly defined. They are sometimes difficult because he writes with odd meters, invented words or phrases, and strange syntax.

Many of Hopkins's poems are paeans to the glory of God and his creations. They begin with lines like "The world is charged with the grandeur of God" and "Glory be to God for dappled things." But others show his apparent despair with his world: "Comforter, where, where is your comforting? / Mary, mother of us, where is your relief?... all / Life death does end and each day dies with sleep." "Mine, O thou lord of life, send my roots rain," an apparent reference to his celibacy.

One of his better known poems is "Spring and Fall: To a Young Child." Here it is. It is fairly difficult, so I have provided a translation in the right column.

Márgarét, are you gríeving
Over Goldengrove unleaving?
Leáves, líke the things of man, you
With your fresh thoughts care for, can you?
Ah! ás the heart grows older

Margaret, do you grieve because the leaves are falling from the trees in Goldengrove? Do you, with your unspoiled thoughts, care for leaves as for human artifacts?
As you get older, you will be

290

It will come to such sights colder	able to look at such sights dispassionately, even in the face of vast piles of dead leaves.
By and by, nor spare a sigh	
Though worlds of wanwood leafmeal lie;	
And yet you wíll weep and know why.	Later, when you weep, you will know why; now the reason is not important.
Now no matter, child, the name:	
Sórrow's spríngs áre the same.	All sorrows have the same origin.
Nor mouth had, no nor mind, expressed	Neither your intellect nor your spirit realizes what your heart has guessed.
What heart heard of, ghost guessed:	
It ís the blight man was born for,	
It is Margaret you mourn for.	You are mourning your own death.

This from a man who chose his religion and eventually became a priest. Yet he is in deep despair. In Hopkins's view, everything can be traced to knowledge of death: Sórrow's spríngs áre the same. Death is the *blight* man was *born for*. It is no wonder that some form of afterlife or reincarnation is common to all religions, whether or not they are God-centered religions.

* * *

I am a scientist, and I accept your argument that processes such as radioactive decay must be deterministic. But I cannot apply it to our brains.

Why not? Don't you think you are a biological system?

Well, yes, but I feel as if I am an independent agent and have free will.

Of course you feel as if you have free will, but do you really? Consider Damasio's patient Elliott (see "Damasio and Descartes," Chapter 7). Elliott invariably made the wrong decisions, even though he eagerly wanted to win the game. He simply could not make the right choice. He had to go for broke. He had no free will.

Now, Elliott had damage to a single part of his brain, as far as we know. Unless you want to argue that he was damaged in precisely the place where the free will resides, you have to recognize that the control group had no more free will than Elliott and the others in the study group.

Many patients with Tourette's syndrome have an apparently irresistible urge to curse. Sometimes it gets them into a lot of trouble, but they cannot stop, any more than you can refrain from scratching an itch. Other patients have other fixations. People with phobias may be unable to enter an elevator or go outdoors. Addicts cannot give up cigarettes, alcohol, or narcotics, sometimes despite desperately wanting to do so. Do they have free will in all respects but those? Or is it merely obvious that those patients lack free will to that extent, and less obvious but equally true of the rest of us?

An engineering professor is building some robots that will work similarly to insects. When one of these robots comes to an obstacle, it will "feel" it with an antenna or "see" it with an eye. It will then "decide" whether to go to the left or the right. Has it free will? A chess-playing computer chooses its next move from a substantially larger array of choices than the robot, and it makes a decision. Has it free will? Or does it merely seem to have free will because of the vast array of options open to it and the complexity of its decision? Have we free will? Or do we merely seem to have free will because of the vast array of options open to us and the complexity of our decisions? I think the latter.

I cannot accept that we are merely biological systems. I believe that we have a spark of life, or a soul, that is nonphysical.

OK. That is a reasonable hypothesis. Can you present evidence to support it? I have presented evidence to support my hypothesis by showing that many of the mechanisms of life can be ascribed to physical and chemical processes, and that the brain seems to work like an organ with very specific functions. Can you support your contention?

I base it on my reading of the Bible: God clearly intends humans to have free will, or else he would not have issued all those commands (through the Prophets, for example) and allow us either to follow them or not.

Why the Bible? Why not the Koran or the Bhagavad-Gita?

The Bible is a historical document, and the life of Jesus is well attested in history. The Gospels are eye-witness accounts written at the time.

Do you not know that most scholars date the earliest Gospel to a time roughly 40 years after the supposed Resurrection? That Luke, Matthew, and John may be based in part on Mark and therefore are not wholly independent accounts? And that we have no contemporary written references to Jesus outside the New Testament?

Josephus?

No. The passage by Josephus (37 or 38-101 C.E.) about Jesus is thought to have been doctored or added by later

293

Christian redactors. Likewise, Tacitus (55?-after 117 C.E.) makes a reference to a Christus who was executed by Pilate, but that is by no means a contemporary account, and it postdates the Gospels. Neither Josephus nor Tacitus had been born at the time of the Crucifixion.

I am distrustful of such so-called "scholars" anyway.
It frankly troubles me how little regard you have for Biblical studies and interpretations other than your own. Many of the scholars I am talking about are mainstream Protestant clergypersons. On what basis do you dismiss their conclusions?

I don't exactly dismiss their conclusions; rather they are of no interest to me. I am not a scholar myself. My Christian [that is, fundamentalist Christian] teachers tell me that the Gospels are eye-witness accounts, and this has been demonstrated in many volumes. On the other hand, you quote people who say that they were written over a generation after the Crucifixion. I have no way to judge, so I maintain my belief.
You can evaluate some of the claims yourself. For example, two of the Gospels give completely different genealogies for Jesus [Matthew 1:1-16; Luke 3:23-38]. That is only one of many contradictions. Are they both correct?

I have heard that one is really a genealogy of Mary.
I have heard that too. However, there is no evidence to support the claim. Rather, it is an *ad hoc* hypothesis designed to patch up the theory.

I do not care if there are small errors in the Bible; these could be errors that propagated as the original, which was truly the word of God, was copied and handed down. I also do not happen to believe in evolution, though I can accept that the first days of creation might not have been literal 24-hour days. (God must have put humus, which is rotted plant matter, into the ground so the first trees could grow; why not fossils?) But I cannot accept the idea that, say, some of the words attributed to Jesus are not really his words. I try to live my life according to the Bible. If it were not the literal word of God, it would be a less clear guide and less valuable to me.

Suppose you found that one of the statements attributed to Jesus was in fact incorrectly attributed. Would that shake your faith in the entire edifice?

Yes; I would not know what was truly the word of God and what was the word of men.

I do not understand why you cannot make your own decisions, but I guess we had better leave that for another discussion.

* * *

Nothing we know of is permanent. We are not permanent; our culture is not permanent; our species is not permanent; neither is the earth or the solar system. Doesn't it create a void in your life to know that everything is in the long run pointless?

I would say, rather, that it exposes a void, not creates it. But I am not willing to fill that void by pretending—I do not believe in Pascal's wager.

Pascal's wager?

Pascal was the founder of modern probability theory. He argued that if he did not believe in God and was wrong, then he would almost surely go to Hell or, at least, not go to Heaven. On the other hand, if he believed in God and was right, then he had at least a chance of Heaven and eternal life. Furthermore, if he is right in believing in God, then the payoff is great, whereas, if he is wrong, the loss is small. Therefore, as a matter of prudence, he would believe in God. (Pascal seems not to have thought of the possibility that some other entity might decide to send him to Hell for believing in God; that possibility reduces his chances of going to Heaven by half.)

295

To me, Pascal's wager is not prudence but hypocrisy. Surely, if Pascal's conception of God is accurate, then God will know that Pascal is pretending. Nor is God so insecure that he needs Pascal's praises for his well-being. God will judge Pascal on Pascal's merits, not on whether he believes in God. People who believe in Pascal's wager think they can play God for a sucker.

If you believe that you have no free will and life is in essence pointless, then why don't you go to New York and jump off the Brooklyn Bridge?

I don't want to. That's an internal reality. Whatever the internal state of my brain is, it enjoys life and wants to survive. Furthermore, that should not be surprising: Natural selection favored organisms that looked out for themselves and feared death. Those that did not fear death were probably reckless and were killed without progeny.

If I have no free will, then why do I have to behave morally?

I don't know that you have to behave morally, but I can suggest that you have no choice but to behave as if you had free will. Otherwise, you would just sit there and starve to death worrying about your loss of free will. Or, as you put it, you would jump off the Brooklyn Bridge.

At any rate, I agree with Stephen Hawking, the well known cosmologist, that free will is a good approximation to human behavior. [Ferguson, 1995] That is so in the same sense that the laws of probability are a good approximation to rolling the dice. Since we cannot develop more precise theories, we must pretend both that we have free will and that the roll of the dice is random. But neither is strictly true.

As to moral behavior, I will suggest further that pain is real and you should not needlessly inflict pain on someone else, simply because it hurts that person. The Golden Rule, if you will. I cannot give you a better reason. But I am certain that,

296

whether or not we have free will, we can surely feel pain, whether emotional or physical.

Aren't you compartmentalizing?
Of course I am. But that is not necessarily pathological. For example, a physician or a social worker has to keep her emotions separate from her work, or else she will drive herself crazy. A well known physician once said, approximately, "I can look with cold dispassion at the body of my friend at an autopsy, but I will weep for him at his funeral." Similarly, when I say, "We are going to be stuck with those reactionaries for generations," I am compartmentalizing in the sense that I know that I will not be around for generations.

I live my life as if I had free will because there is no other choice, but intellectually I do not believe that it is truly free will. And I am unwilling to say, with William James, that my first act of free will will be to decide that I have free will. That's cute, but it's begging the question. I prefer to find out the truth, at all costs.

Why do we punish criminals if they have no free will?
We live in a society with other people. As a practical matter, if not as a matter of morality, we cannot have our citizens stealing from each other or murdering each other. We have to have a means of control. We can train most people not to rob or murder when they are children. Those we cannot so train, we have to control in some other way: either punishment or segregation from the rest of society.

In an interview, Will Provine, a biologist from Cornell University, expressed views similar to mine; he argues that a chess-playing computer, for example, makes decisions, but that is not evidence of free will. [Stannard, 1996] He argues that morality is necessary to live in a society but that it is instilled by conditioning. Provine regards free will as an excuse society uses to punish transgressors harshly. He thinks, rather, that prison sentences or other punishment should be considered an

297

opportunity for reconditioning; when a person is clearly reconditioned and can be a productive member of society, Provine says, he should be released. I am in substantial agreement with this sentiment, though I add that some people cannot be reconditioned or can fool us into thinking they are reconditioned. In addition, it is possible that long sentences for some transgressors might deter others from transgressing, though that is not obvious.

So that's why we have to be ethical? Because society cannot allow us to be otherwise?

No. As I implied above, I think ethics means not unnecessarily harming other people, and I think we should not unnecessarily harm other people. We sometimes have to punish people if they are sufficiently unethical, that is, if they unnecessarily harm other people.

And ethics does not follow from religion?

No, though it may be true that our ethical view of the world was originally developed by religious people. Nevertheless, religion has too often been unethical to lay exclusive claim to ethical behavior.

I would like to let Harold Kushner have the last word. In an interview with the *Jerusalem Post*, [Gillon, 1985] Kushner outlined his belief that God could not be omnipotent. He compared someone who worships an omnipotent God with someone who worships power. I have found much fault with Kushner's reasoning (see "Reward and Punishment," Chapter 3), but here I think he is exactly correct: You don't have to believe in God to be religious in a very broad sense; you have only to behave in a certain way. As Kushner says,

> I even think that a person who believes in love and compassion, and practices those things, even though he says he does not worship God, is more religious than the worshipper of power.

298

The study of the behavior of subatomic particles in this century is supposed to have established at least three exceedingly curious facts about the physical world.

DAVID Z ALBERT

Appendix
Random determinism

Schrödinger's cat

The physicist Erwin Schrödinger, one of the founders of quantum mechanics, proposed a thought experiment now called the Schrödinger's cat experiment. In this thought experiment, we place a cat in a box with a radioactive material such as radium. Let us for simplicity choose the amount of radium so that the odds are fifty-fifty that a single atom will decay after one day. If one atom decays, the alpha particle it emits is detected by a Geiger counter. An electronic circuit releases a weight that breaks a vial of cyanide and kills the cat. If no atom decays, we expect to find the cat alive and well at the end of the experiment. [Gribbin, 1984]

At the end of one day, the cat is either dead or alive. We do not know which and express our uncertainty by stating that the probability that the cat is alive is 1/2, and the probability that the cat is dead is 1/2. When we open the box, we find out which condition is true, and one of the probabilities instantly changes to 1 and the other to 0. This happens because we are now certain, whereas before we were uncertain.

Let us express the condition that the cat is alive as ☺ and the condition that the cat is dead as ☹. In quantum mechanics, we use a mathematical entity called a *wavefunction* to describe the state of a particle, a collection of particles, or a cat. The wavefunction is expressed symbolically with this kind of

299

bracket: | >. It is known as a *ket*, a word coined by Dirac because it is the right half of a brac*ket* like this one: < | >. (Dirac called the left half < | a *bra*, incidentally, but he did so before the word had acquired its more common use.) The wavefunction for a living cat is |☺>, whereas the wavefunction for a dead cat is |☹>. At the outset of the experiment, the cat is surely alive, so the wavefunction is |☺>. After a very long time, the cat will surely be dead, so the wavefunction will be |☹>. After one day, when his odds are fifty-fifty, the cat is said to be in a *superposition of states*, and his wavefunction is (|☺> + |☹>)/√2. That is, the wavefunction evolves with time in a very definite way. The factor of 1/√2 is a normalizing factor designed to make the sum of the probabilities equal to 1. Its value is of no importance to our argument.

When we open the box after one day, the wavefunction (|☺> + |☹>)/√2 immediately changes to |☺> or |☹> because we know the state of the cat. The wavefunction can change instantly in that way because, at least in this case, it has no physical significance other than as a certain kind of probability function. (Technically, the absolute square of the wavefunction, not the wavefunction itself, is the probability.) I will argue that the wavefunction of the cat either remains |☺> or changes to |☹>, and has already done so before we open the box. The statement that the wavefunction is (|☺> + |☹>)/√2, therefore, is really a statement about our knowledge of the wavefunction, not a statement about the actual wavefunction or the state of the cat. It is easy to confuse our knowledge about a system with the state of the system itself.

Trying to give too much significance to the wavefunction has led a great many, to believe, for example, that the cat is neither dead nor alive until we open the box. [Gribbin, 1984] Then, if the cat turns out to be alive here, another universe in which the cat is dead is instantly created. That is, Gribbin argues that a quantum process is never completed until it is observed by a conscious being. (It is difficult to know what counts as a conscious being, and Gribbin discusses this point, but for discussions of the cat paradox, the cat is usually ruled out. A television camera or a photodetector does not qualify as an observer because it is itself a quantum system. It is never made clear why a conscious observer is not also a quantum system, nor how the experiment would be different if the cat were replaced by a conscious being such as a human.) Only when the outcome of the process is known does the wavefunction collapse to a definite value; then a parallel universe is created instantly. I have no idea where the energy for creating that universe is supposed to come from. The Schrödinger cat experiment convinces me, to the contrary, that whatever happens inside the box, happens; we find out what when we look.

Let me propose another thought experiment, though this is one that we could easily perform if we wanted to. It is the classical double-slit experiment. We can perform it with light or with electrons; because of wave-particle duality, the experiments are in principle the same. Let us assume, however, that we do the experiment with light. [Greene 2000, Chap. 4]

In the classical double-slit experiment, we illuminate two narrow slits with a beam of light and place a ground-glass viewing screen a distance away. Because of the wave nature of light, we do not see the shadows of the two slits (more accurately, the shadow of the material into which the slits are cut) but rather a series of bright and dark bands called an *interference pattern*. The interference pattern comes about because the light propagates as a wave. The waves that pass through the slits travel different distances to different parts of the viewing screen. When the distance is such that the waves are out of phase with one another (one wave waving up while the other waves down, for example), we see a dark band, whereas we see an especially bright band when the waves arrive at the screen in phase (both waving up, for example). For over a century, this experiment was regarded as proof that light was a wave.

Let us now replace the ground-glass screen with a photographic plate. A photographic plate consists of a

301

suspension of light-sensitive *grains* in a thin gelatin layer that is attached to a base such as glass. For simplicity, let us say that a grain becomes *developable* when it absorbs a single particle of light, or *photon*. That is, that grain can be made opaque by chemical processing, whereas other, unexposed grains will not become opaque after the same processing. After enough grains have been made developable, we can develop the film and see the double-slit pattern.

Suppose further that we reduce the intensity of the light until the photons fall one at a time onto the plate. Then, photons will be absorbed sporadically around the plate. Each such absorption is a quantum event, and there could be millions of them on a single plate. None, however, is actually observed until we develop the plate and look at it. According to the alternative universe hypothesis, as soon as I take the plate out of the developer and look at it, millions of alternative universes are created, at least one for each grain on the plate. Before then, each grain was in a mixed quantum state, like the cat before we opened the box.

I could have waited years before I developed the plate. Until then, the statement that each grain is in a mixed quantum state was only a way of saying that I did not know which grains had been made developable and which had not. In fact, certain grains had been made developable and remained in that state until the film was chemically processed. The processing simply revealed which grains had been made developable and which had not. There is no need to imagine an alternative universe for each grain nor to imagine some sort of mystical significance for the role of the observer.

Requiring a conscious observer to record a quantum process before the process is deemed real or complete is not an idiosyncratic invention of Gribbin. Many physicists have worried about the role of the observer in quantum mechanics. Partly, this is necessary because the fact of observation often disturbs the system enough to destroy an experiment. For example, if we tried to perform the double-slit experiment in

302

such a way as to determine precisely which slit the particle went through, we would destroy the interference pattern entirely. That is, in quantum mechanics, it is impossible to track a particle through an apparatus and still perform the intended experiment; tracking the particle exerts a force on it and inevitably changes the outcome of the experiment. Many have used this fact to conclude that there is no real, objective world and we cannot even talk about what a particle does between the time it leaves the source and reaches the detector. I do not know anyone who really believes this interpretation, and the mathematician Sheldon Goldstein, [1998] perhaps exaggerating a bit for effect, doubts that anyone "really ever did." Gribbin nonetheless concludes that a particle does not really exist or that a quantum interaction does not really take place until it is observed by a consciousness.

To accept this conclusion, which many well-known physicists have either shared or toyed with, you have to treat the observer as a classical (nonquantum) system, and there is no justification for doing so. More probably, the observer himself is a quantum mechanical system. The fact of observation is therefore a quantum mechanical event inside his brain. Who observes the observer to ensure that the observation has really taken place? Who observes that observer? We are left with another infinite regression.

What is so "weird" about quantum mechanics that it makes people worry about the role of the observer and propose alternative universes? Suppose that I had replaced the ground-glass screen with a phosphorescent screen instead. This is a screen that is coated with material that absorbs a photon and almost immediately emits another photon with a slightly lower energy. If we watch, we will see flashes of light sporadically around the screen. If we record the locations of the flashes, we will eventually build up the double-slit interference pattern.

These flashes represent a very localized interaction. The photons interact with the atoms in the screen as if photons are particles. Indeed, the photoelectric effect, where a photon drives

303

an electron (not another photon) out of a solid, has long been considered evidence that light is a particle, not a wave.

Here, then, is the difficulty: Light seems to travel through space as if it is a wave, or else there can have been no interference pattern. But when light is absorbed, it behaves as a particle. This is called *wave-particle duality*. Most physicists compartmentalize and use the wave picture when it is convenient and the particle picture when it is convenient.

Quantum mechanics demystified (a little)

How can we explain wave-particle duality? Many physicists refuse to think about it and instead adopt the instrumentalist's viewpoint: We have a theory and it works; never mind why. This is not exactly a religious statement, because the theory is empirically based and falsifiable; indeed, it has survived countless attempts at falsification. Nevertheless, these physicists consider an electron or a photon as a particle or as a wave, as is convenient. For propagation experiments, such as the double-slit experiment, they use the wave picture, whereas they use the particle picture for interactions such as phosphoresence. They dismiss as mere philosophy all attempts to understand quantum mechanics from an intuitive perspective. I argue, to the contrary, that we do not understand something well until we have both a mathematical and an intuitive or descriptive argument; neither alone is enough.

Sometimes it is necessary to rush in where Nobel laureates fear to tread. Let us begin by recognizing that *what we call free space has structure*.

Newton was greatly troubled by his own theory of gravity because it required *action at a distance*. That is, it required a body, such as the sun, to exert a force on another body, such as a planet, without any apparent mechanism that could have created that force. This violates a principle that the physicist Max Born, [1964] another of the founders of quantum theory, calls *contiguity*. Contiguity is the postulate that cause and effect must

304

be in physical contact or at least connected by a chain of intermediate causes that are themselves in physical contact. Roughly speaking, you cannot move an object unless you touch it or touch it with something else. Newton's theory apparently violated the principle of contiguity but nevertheless worked, though for several centuries it was necessary to adopt the instrumentalist's viewpoint and simply accept it in spite of the troubling action at a distance.

Einstein's General Theory of Relativity, however, shows that a mass such as the sun curves the space around it so that another mass such as the earth behaves precisely as if it were attracted by a Newtonian force. In this sense, free space must have some kind of structure. Since the universe consists of a great many masses always in constant motion with respect to each other, I visualize that structure as rough and constantly changing on the submicroscopic scale.

What about a quantum-mechanical particle? I imagine a moving particle as a hard core that moves without resistance through free space, as if free space were a sort of superfluid. Associated with the particle is a wake, similar to the wake of a boat on the surface of the water. The wake represents the long-range field of the particle. It is an integral part of the particle in that you can never separate the hard core from the wake. The propagation of the particle through space is governed almost entirely by the wake. In the double-slit experiment, the hard core of the particle goes through one slit or the other, but you cannot determine which because your attempt to do so destroys the experiment. When the particle arrives at the film plane, however, it is presumably the hard core that interacts at a discrete location with one of the grains in the emulsion. The wake then goes out of existence; with Philip Wallace [1996] of McGill University, I assume that it does not disappear instantaneously but rather decays away gradually at all locations. This view gives physical significance to the wavefunction, and Wallace argues that the field is all that is real. I cannot go quite so far, in part because Wallace's view cannot provide a concrete

305

model that describes why the interactions always occur at isolated points.

Thus, my intuitive picture of quantum mechanics is this: I postulate that the path of the hard core is completely determined by the initial direction and velocity (the initial conditions) but that the path is unpredictable because the hard core is constantly buffeted by random forces similar to choppiness on the surface of the sea. These forces arise because free space is rough as the result of forces due to all the other particles in the universe; the roughness constantly changes as those particles move with respect to our experiment. Thus, the hard core can end up anywhere in the film plane but, on average, grains will be made developable where predicted by wave theory. That is, quantum mechanics cannot predict the path of an individual particle but yields only the average path of a large number of particles. The wake is not measurably affected by the choppiness because the wake represents a long-range part of the particle, whereas the choppiness is described by very short-range fluctuations that average to zero over the scale of the wake.

For experts: It is hard to reconcile this point of view with the Einstein-Podolsky-Rosen experiments (and, indeed, any experiments that involve spin or polarization), but I do not think that these experiments are wholly understood. (In retrospect, however, the results of the EPR experiments should not have been surprising; we know very well that a particle's influence can extend for a macroscopic distance. What was surprising was perhaps the magnitude of that distance.) First, the EPR experiments leave us in a position similar to that of Newton in that we have an experiment that we cannot understand unless we invoke action at a distance. I think we simply have to wait until someone explains them, as Einstein explained the origin of gravity. Second, in experiments, such as the Stern-Gerlach experiment, that separate particles according to spin or polarization, I have always found it mysterious that *we* get to choose the z axis by how we orient our magnet or our polarizer. I suspect that we will not understand the EPR experiments until

306

we understand quantum mechanically what happens when a photon strikes a polarizing beam splitter or when an electron enters a Stern-Gerlach magnet. I do not deny quantum mechanics but rather argue that there is probably some underlying theory, more fundamental than quantum mechanics, which we have not yet formulated. Following Laplace, Einstein, and Bohm, I postulate that this theory will be deterministic.

This view is not inconsistent with a theoretical development of David Bohm and Basil Hiley, and later researchers. [BBC, 1962; Davies and Brown, 1986, chaps. 8 and 9; Hiley and Peat, 1987; Albert, 1994; Lindley, 1996. The mathematician Sheldon Goldstein [1998] outlines several formulations of what he calls quantum mechanics without observers; Bohm's is one of these.] Bohm's theory, which is a hidden variable theory, has so far produced no testable differences from quantum mechanics and also requires action at a distance. It invokes a potential that travels faster than light, but it is not clear whether the stricture against propagation faster than light holds in quantum mechanics. Relativity is a classical theory in the sense that it is non–quantum mechanical, and it has so far proved impossible to integrate it with quantum mechanics. Weinberg [1994] notes that the EPR experiments do not allow *signals* to be sent faster than light, and relativity does not forbid *correlations* or *fields* to propagate faster than light. He expresses sympathy for the realist viewpoint, even at the expense of allowing action at a distance.

Finally, I think the problem of the observer can be finessed by noting that some quantum interactions are, in effect, irreversible. That is, they have an immeasurably small probability of being reversed. Thus, a photon can flutter around (figuratively speaking) changing from one quantum state to another and back for a long time, but sometimes it undergoes an interaction such as absorption. The probability that an electron in a trap in a photographic grain will decay and emit a photon into the same quantum state as the incoming state is negligible, so we may say that the photon was absorbed whether or not

307

anyone was present to observe it. The relatively new concept of decoherence also addresses this point. [Lindley,1996]

What about Schrödinger's cat? We greatly oversimplify when we describe the wavefunction of the cat as a superposition of two states, $|☺>$ and $|☹>$. In fact, the wavefunction of the cat is an enormously complicated mathematical function that would contain more variables than the number of atoms in the cat. That wavefunction evolves with time. If the radium atom decays, then cyanide is released into the air. That cyanide binds in an irreversible quantum mechanical reaction with the hemoglobin in the cat's blood and prevents the cat's tissues from getting oxygen. The cat stops breathing, and his bodily functions slow and stop. You cannot perhaps pinpoint the exact moment when the cat dies, but die he does. We find out his fate when we open the box. There is no other, living cat somewhere else in another universe.

References

Albert, David Z, 1994, "Bohm's Alternative to Quantum Mechanics," *Scientific American*, May, pp. 58-67.

Allen, Steve, 1990, *Steve Allen on the Bible, Religion, and Morality*, Prometheus.

Allen, Steve, 1993, *More Steve Allen on the Bible, Religion, and Morality*, Prometheus.

Anderson, Bernhard W., 1966, *Understanding the Old Testament*, Prentice-Hall.

Asimov, Isaac, 1981, *Asimov's Guide to the Bible*, Avenel.

Barbour, Ian G., 1974, *Myths, Models and Paradigms, A Comparative Study in Science and Religion*, Harper & Row.

Barbour, Ian G., 1990, *Religion in an Age of Science: The Gifford Lectures, 1989-1991, Volume 1*, Harper & Row; expanded and reissued as Barbour, Ian G., 1997, *Religion and Science: Historical and Contemporary Issues*, HarperCollins.

Bar-Hillel, Maya, Dror Bar-Natan, and Brendan McKay, 1997, "There Are Codes in *War and Peace* Too," *Galileo*, November-December (in Hebrew). I thank Maya Bar-Hillel of the Hebrew University for supplying me with an English translation of this article.

Ben-David, Calev, 1995, "Who Wrote the Bible?" *Jerusalem Report*, 15 June, pp. 20-26. See also letter to the editor by Jacob Mendlovic, 10 August, p. 3.

Blackmore, Susan, 1986, *The Adventures of a Parapsychologist*, Prometheus.

Blackmore, Susan, 1991, "Near-Death Experiences: In or Out of the Body," *Skeptical Inquirer*, Fall, pp. 34-45.

Blackmore, Susan, 1996, "Near-Death Exepriences," in Gordon Stein, ed., *The Encyclopedia of the Paranormal*, Prometheus, pp. 425-441. See also "Out-of-Body Experiences," pp. 471-483.

Blackmore, Susan, 1998, "Abduction by Aliens or Sleep Paralysis?" *Skeptical Inquirer*, May/June, pp. 23-28.

Block, J. R., and Harold Yuker, 1989, *Can You Believe Your Eyes? Over 250 Illusions and Other Visual Oddities,* Gardner.

Bohm, David, and Basil J. Hiley, 1995, *The Undivided Universe,* Chs. 1, 2, Routledge.

Bohm, David, and F. David Peat, 1987, *Science, Order, and Creativity,* Bantam.

Born, Max, 1964, *Natural Philosophy of Cause and Chance,* Dover.

Born, Max, 1956, *Experiment and Theory in Physics,* Dover.

Bower, Bruce, 1996, "Remembrance of Things False," *Science News,* 24 August, pp. 126-127.

Bragg, Rick, 1994, "Tried by Deadly Tornado, An Anchor of Faith Holds," *The New York Times,* 3 April, pp. 1A, 12A.

British Broadcasting Corporation, 1962, *Quanta and Reality,* American Research Council.

Bronowski, J., 1973, *The Ascent of Man,* Little, Brown.

Brooks, Anne Marie, 1993, "Gateway Drugs," *Current Health,* January, pp. 6-11.

Carter, Rita, 1996, "Holistic Hazards," *New Scientist,* 13 July, pp. 12-13.

Carter, Dan, 1992, "The Academy's Crisis of Belief," *The Chronicle of Higher Education,* 18 November, p. A36.

Casti, John L., 1989, *Paradigms Lost: Images of Man in the Mirror of Science,* Avon.

Chomsky, William, 1957, *Hebrew: The Eternal Language,* Jewish Publication Society of America.

Cohen, Sharon, 1995, "Nurse Still Gave Life – Even in Death," Boulder *Daily Camera,* 26 April, pp. 1A, 7A. See also 22 April, 1995, p. 8A.

Concar, David, 1998, You Are Feeling Very, Very Sleepy...," *New Scientist,* 4 July, pp. 26-31.

Consumer Reports, 1994, "Homeopathy: Much Ado about Nothing?" March, pp. 201-206.

CQ Researcher, 1995, "Is Marijuana a 'Gateway' Drug that Leads Users to Try More Dangerous Drugs like Cocaine?" 28 July, p. 673.

Crews, Frederick, 1996, *The Memory Wars: Freud's Legacy in Dispute*, The New York Review of Books.

Damasio, Antonio R., 1995, *Descartes' Error: Emotion, Reason, and the Human Brain,* Avon.

Damasio, Antonio R., and Hanna Damasio, 1992, "Brain and Language," *Scientific American*, September, pp. 89-95.

Davies, Paul, 1992, *The Mind of God: The Scientific Basis for a Rational World*, Touchstone.

Davies, P.C.W., and J.R. Brown, 1986, *The Ghost in the Atom: A Discussion of the Mysteries of Quantum Physics*, Cambridge University Press.

Dawkins, Richard, 1995, *River out of Eden*, Basic.

Dawkins, Richard, 1987, *The Blind Watchmaker*, Norton.

Dawkins, Richard, 1989, *The Selfish Gene* (new edition with notes by the author), Oxford University Press.

de Waal, Frans, 1996, *Good Natured, The Origins of Right and Wrong in Humans and Other Animals*, Harvard University Press.

de Duve, Christian, 1996, "The Birth of Complex Cells," *Scientific American*, April, pp. 50-57.

Dean, Geoffrey, 1987, "Does Astrology Need to Be True? Part 2: The Answer Is No," *The Skeptical Inquirer*, Spring, pp. 257-273.

Dean, Geoffrey, 1986-1987, "Does Astrology Need to Be True? Part 1: A Look at the Real Thing," *The Skeptical Inquirer*, Winter, pp. 166-184.

Dennett, Daniel, 1995, *Darwin's Dangerous Idea*, Simon and Schuster.

Dennett, Daniel, 1996, *Kinds of Minds*, BasicBooks.

Diet and Nutrition Newsletter, 1994, "Sugar's Link to Behavior: A Sweet Tale after All," April, Tufts University, pp. 1-2.

Drosnin, Michael, 1997a, *The Bible Code*, Simon and Schuster.

311

Drosnin, Michael, 1997b, "Battling over the Codes" (letter to the editor), *Jerusalem Report*, 2 October, p. 2.

Ebbern, Hayden, Sean Mulligan, and Barry L. Beyerstein, 1996, "Maria's Near-Death Experience: Waiting for the Other Shoe to Drop," *Skeptical Inquirer*, July-August, pp. 27-33.

Einstein, Albert, 1954, *Ideas and Opinions,* Crown.

Eisenstein, Ira, 1964, *What We Mean by Religion* (revised and enlarged edition), Reconstructionist Press.

Ehrenreich, Barbara, and Janet McIntosh, 1997, "The New Creationism," *The Nation*, 9 June, pp. 11-16.

Ferguson, Kitty, 1995, *The Fire in the Equations: Science, Religion, and the Search for God*, Eerdmans.

Ferris, Timothy, 1995, "Post-Modernist Piffle" (letter to the editor), *The New York Times Book Review*, 15 October.

Ferris, Timothy, 1997, *The Whole Shebang, A State of the Universe(s) Report*, Simon and Schuster.

Fischman, Joshua, 1995, "Why Mammal Ears Went on the Move," *Science*, 1 December, p. 1436.

Flesch, Rudolf, 1951, *The Art of Clear Thinking*, Harper & Row.

Folger, Tim, 1997, "Cosmic Evolution" (review of Smolin, 1997), *Discover*, July, pp. 120-121.

Freke, Timothy, and Peter Gandy, 2000, *The Jesus Mysteries: Was the "Original Jesus" a Pagan God?*, Harmony Books.

Friedlander, Michael W., 1995, *At the Fringes of Science*, Westview.

Friedman, Richard Elliott, 1987, *Who Wrote the Bible?* Summit.

Gallagher, Winifred, 1994, "How We Become What We Are," September, pp. 39-55.

Gardner, Martin, 1989, *Science: Good, Bad and Bogus,* Prometheus.

Garnett, Leah R., 1995, "Is Less Really More?" *Harvard Health Letter*, May, pp. 1-3.

Garraty, John A., and Peter Gay, 1972, *The Columbia History of the World*, Harper & Row.

Gazzaniga, Michael, 1995, "Gut Thinking" (book review), *Natural History*, February, pp. 68-71.

Gibbons, Ann, 1996, "On the Many Origins of Species," *Science*, 13 September, pp. 1496-1499.

Gillon, Philip, 1985, "When a Bad Thing Happened to a Good Person...," *The Jerusalem Post International Edition*, 24 August, p. 14.

Glatzer, Nahum N., 1969, *The Dimensions of Job – A Study and Selected Readings*, Schocken.

Glynn, Patrick, 1997, *God: The Evidence: The Reconciliation of Faith and Reason in a Postsecular World*, Prima.

Goldstein, Sheldon, 1998, "Quantum Theory without Observers – Part One," *Physics Today*, March, pp. 42-46; "Quantum Theory without Observers – Part Two," *Physics Today*, April, pp. 38-42.

Goodwin, Brian, 1994, *How the Leopard Changed Its Spots: The Evolution of Complexity*, Scribner.

Gould, Stephen Jay, 1993, "An Earful of Jaw," Ch. 6 of *Eight Little Piggies*, Norton.

Gould, Stephen Jay, 1994, "The Evolution of Life on the Earth," *Scientific American*, October, pp. 85-91.

Gould, Stephen Jay, 1991, "Life's Little Joke," Ch. 11 of *Bully for Brontosaurus*, Norton.

Gould, Stephen Jay, 1991, *Wonderful Life: The Burgess Shale and the Nature of History*, Penguin.

Graves, Robert, and Raphael Patai, 1983, *Hebrew Myths: The Book of Genesis*, Ch. 1, Greenwich House.

Greenberg, Joanne, 1964, *I Never Promised You a Rose Garden*, Holt, Rinehart, and Winston (originally published under the pseudonym Hannah Green).

Greene, Brian, 2000, *The Elegant Universe*, Vintage.

Gribbin, John, 1984, *In Search of Schrödinger's Cat: Quantum Physics and Reality*, Bantam.

Gross, Paul R., and Norman Levitt, 1994, *Higher Superstition: The Academic Left and Its Quarrels with Science*, Johns Hopkins University Press.

Harris, Lis, 1986, *Holy Days: The World of a Hassidic Family*, Collier/Macmillan.

Harvard Health Letter, 1994, "Peptic Ulcer Disease: Bugging the System," June, pp. 1-3. See also "The Ulcer Germ: Stomach Cancer Too?" October, 1997, pp. 6-7.

Henarejos, Philippe, 1999, "La Nouvelle Guerre des Étoiles (The New Star Wars," *Science et Vie*, Juin, pp. 72-77 (in French).

Hendel, Ronald S., 1997, "The Secret Code Hoax," *Bible Review*, August, pp. 23-25. See also Anonymous, 1997, "The Bible Code: Cracked and Crumbling," *Bible Review*, August, p. 22.

Hick, John, 1964, *The Existence of God*, Macmillan.

Hiley, Basil J., and F. David Peat, 1987, *Quantum Implications: Essays in Honour of David Bohm*, Chs. 1, 2, 5, 8, 9, 12, Routledge.

Holden, Constance, 1980, "Twins Reunited," *Science 80*, November, pp. 55-59.

Horgan, John, 1995, "The New Social Darwinists," *Scientific American*, October, pp. 174-181.

Horovitz, David, 1997, "Busting the Bible Code Breakers," *Jerusalem Report*, 4 September, pp. 14-21.

Hull, David L., Peter D. Tessner, and Arthur M. Diamond, 1978, "Planck's Principle: Do Younger Scientists Accept New Scientific Ideas with Greater Alacrity than Older Scientists?" *Science*, 17 November, pp. 717-723.

Hume, David, 1990, *Dialogues Concerning Natural Religion*, Martin Bell, ed., Penguin.

Huston, Peter, 1992, "Night Terrors, Sleep Paralysis, and Devil Stricken Demonic Telephone Cords from Hell," *Skeptical Inquirer*, Fall, pp. 64-69.

Kandel, Minouche, and Eric Kandel, 1994, "Flights of Memory," *Discover*, May, pp. 32-38.

Kaplan, Mordecai, 1967, "The Past Stages of the Jewish Religion," Ch. 25 of *Judaism as a Civilization*, Schocken.

Kelemen, Lawrence, 1990, *Permission to Believe, Four Rational Approach to God's Existence*, Targum/Feldheim.

Kee, Howard Clark, Franklin W. Young, and Karlfried Froehlich, 1965, *Understanding the New Testament*, 2nd ed., Prentice Hall.

Kerr, Richard A., 1995, "A Volcanic Crisis for Ancient Life?" *Science*, 6 October, pp. 27-28.

Kolata, Gina, 1983, "Dietary Dogma Disproved," *Science,* 29 April, pp. 487-488. See also O'Dea, Kerin, 1983, "Diabetes and Diet" (letter to the editor), 15 July, p. 214.

Kuhn, Thomas, 1970, *The Structure of Scientific Revolutions* (2nd ed., enlarged), University of Chicago Press.

Kushner, Harold S., 1981, *When Bad Things Happen to Good People*, Schocken.

Kushner, Harold S., 1976, *When Children Ask about God,* Schocken.

Lane Fox, Robin, 1991, *The Unauthorized Version: Truth and Fiction in the Bible,* Knopf.

Lasky, Ronald C., 1999, *Beyond Reasonable Doubt: Evidence for a Designed Universe*, 1stBooks Library, www.1stbooks.com.

Lawren, Bill, 1992, "The Case of the Ghost Molecules," *Omni*, June, pp. 50-52, 73-74.

Levi, Barbara G., 1973, "Anomalous Water: An End to the Anomaly," *Physics Today*, October, pp. 19-20. See also "Anomalous Water: Polywater or Impurities?" *Physics Today*, October, 1970, pp. 17-20.

Lévy-Leblond, Jean-Marc, 1990, "Did the Big Bang Begin?" *American Journal of Physics*, vol. 58, pp. 156-159.

Lewin, Roger, 1995, "Evolution's New Heretics," *Natural History*, May, pp. 12-17.

Lindley, David, 1996, *Where Does the Weirdness Go? Why Quantum Mechanics Is Strange, but Not as Strange as You Think,* Basic Books.

Loftus, Elizabeth, 1995, "Remembering Dangerously," *Skeptical Inquirer*, March-April, pp. 20-29.

Lucretius, 1951, *The Nature of the Universe,* translated and with an introduction by Ronald Latham, Penguin.

315

Mach, Ernst, 1960, *The Science of Mechanics*, sixth American edition, translated by Thomas J. McCormack, Open Court.

MacLeish, Archibald, 1961, *J.B.*, Sentry.

Marcus, Ralph, 1956, "The Hellenistic Age," In *Great Ages and Ideas of the Jewish People*, Leo W. Schwarz, ed., Modern Library.

Marshall, Eliot, 1980, "Psychotherapy Works, but for Whom?" *Science*, 1 February, pp. 506-508.

Mendillo, Michael, and Richard Hart, 1974, "Resonances," *Physics Today*, February, p. 73.

Michaels, Anne, 1997, *Fugitive Pieces*, Knopf.

Miles, Jack, 1995, *God, a Biography*, Vintage.

Miller, Walter M., Jr., 1955, *A Canticle for Leibowitz*, Matson.

Moreland, J.P., and Kai Nielsen, 1993, *Does God Exist? The Debate Between Theists and Atheists,* Prometheus.

Morrell, Virginia, 1993, "Anthropology: Nature-Culture Battleground," *Science*, 24 September, pp. 1798-1802.

Morris, Thomas V., ed., 1994, *God and the Philosophers: The Reconciliation of Faith and Reason*, Oxford University Press.

Moyers, Bill, 1996, *Genesis*, 31 October, Public Broadcasting System.

Napier, Tom, 1998, "Fundamentally Flawed" (letter to the editor), *Skeptical Briefs*, December.

Neill, A.S., 1966, *Freedom – Not License*, Hart Publishing Company.

New York Times, 1995, "Colleagues Defend Doctor Who Cut Off Wrong Leg," 17 September.

Nickell, Joe, 1998, "Alien Abductions as Sleep-Related Phenomena," *Skeptical Inquirer*, May/June, pp. 16-18.

Nilsson, Dan-E., and Susanne Pelger, 1994, "A Pessimistic Estimate of the Time Required for an Eye to Develop," *Proceedings of the Royal Society of London*, vol. 256, pp. 53-58.

Nyad, Diana, 1995, on National Public Radio, 28 November.

Odenheimer, Micha, 1997, "False Positive, There's a Price to Be Paid for Wanting to Be Proven Right," *Jerusalem Report*, 16 October, p. 59.

O'Neill, Barry, 1994, "The History of a Hoax," *The New York Times Magazine*, March 6, pp. 46-48.

Oliwenstein, Lori, 1993, "Onward and Upward," *Discover*, June, pp. 22-23.

Parrinder, Geoffrey, 1984, *World Religions from Ancient History to the Present*, Facts on File Publications.

Peacocke, Arthur, 1993, *Theology for a Scientific Age: Being and Becoming – Natural, Divine, and Human*, enlarged edition, Fortress.

Pellegrino, Charles, 1995, *Return to Sodom and Gomorrah*, Avon.

Physics Today, 1994, November, p. 115 (cartoon).

Piercy, Marge, 1991, *He, She and It*, Knopf.

Pojman, Louis P., 1987, *Philosophy of Religion*, Wadsworth.

Polkinghorne, John, 1994, *The Faith of a Physicist*, Princeton University Press.

Popper, Karl R., 1968, *Conjectures and Refutations: The Growth of Scientific Knowledge*, Harper Torchbooks.

Potok, Chaim, 1982, "The Bible's Inspired Art," *The New York Times Magazine*, 3 October, pp. 61-68.

Prager, Dennis, and Joseph Telushkin, 1986, *The Nine Questions People Ask about Judaism*, Touchstone.

Ravage, Barbara, 1994, "Marijuana Update," *Current Health*, October, pp. 2-12.

Ross, Hugh, 1998, *The Genesis Question: Scientific Advances and the Accuracy of Genesis*, Navpress.

Sacks, Oliver, 1996, *The Island of the Colorblind*, Vintage

Sagan, Carl, 1978, *The Dragons of Eden: Speculations on the Evolution of Human Intelligence*, Ballantine.

Sagan, Carl, 1986, *Contact*, Pocket Books.

Sagan, Carl, 1997, *The Demon-Haunted World: Science as a Candle in the Dark*, Random House.

Satinover, Jeffrey, 1995, "Divine Authorship? Computer Reveals Startling Word Patterns," *Bible Review*, October, pp. 28-31, 44-45.

Schonfield, Hugh, 1967, *The Passover Plot,* Bantam.

Schroeder, Gerald L., 1992, *Genesis and the Big Bang*, Bantam.

Schroeder, Gerald L., 1997, *The Science of God*, Free Press.

Skinner, B.F., 1976, *Beyond Freedom and Dignity*, Knopf.

Slack, Gordy, 1997, "When Science and Religion Collide, or Why Einstein Wasn't an Atheist," *Mother Jones*, November/December.

Smolin, Lee, 1997, *The Life of the Cosmos*, Oxford.

Sperling, Susan, 1994, "Beating a Dead Monkey," *The Nation*, November 28, pp. 662-665.

Spong, John Shelby, 1991, *Rescuing the Bible from Fundamentalism,* HarperCollins.

Spong, John Shelby, 1994, *Resurrection: Myth or Reality?* HarperCollins.

Stannard, David E., 1980, *Shrinking History*, Oxford University Press.

Stannard, Russell, 1996, *Science and Wonders: Conversations about Science and Belief,* Faber and Faber.

Sternberg, Shlomo, 1997, "Snake Oil for Sale," *Bible Review*, August, p. 24; see also Anonymous, 1997, "The Bible Code: Cracked and Crumbling," *Bible Review*, August, p. 22.

Stove, David, 1992, "The Demons and Dr. Dawkins," *Americam Scientist*, Winter, pp. 67-78.

Swinburne, Richard, 1996, *Is There a God?*, Oxford University Press.

Taylor, John, Raymond A. Eve, and Francis B. Harrold, 1995, "Why Creationists Don't Go to Psychic Fairs: Differential Sources of Pseudoscientific Beliefs," *Skeptical Inquirer*, November-December, pp. 23-28.

Thomas, David E., "Hidden Messages and the Bible Code," *Skeptical Inquirer*, November/December, pp. 30-36. I am further indebted to David Thomas for his talk, "The Bible Code," Rocky Mountain Skeptics, January 17, 1998, and to others who spoke at the meeting for passing along their insights.

Thornhill, Nancy Wilmsen, 1993, *The Natural History of Inbreeding and Outbreeding*, University of Chicago Press.

Tiger, Lionel, and Joseph Shepher, 1975, *Women in the Kibbutz*, Harcourt Brace Jovanovich, p. 7

Tymoczko, Dmitri, 1996, "The Nitrous Oxide Philosopher," *The Atlantic Monthly*, May, pp. 93-101.

Tyson, Neil de Grasse, 1996-1997, "In Defense of the Big Bang," *Natural History*, December-January, pp. 76-81.

Tyson, Neil de Grasse, 1998, "Certain Uncertainties," *Natural History*, October, pp. 86-88; "Belly up to the Bar," *Natural History*, November, pp. 70-74.

UC Berkeley Wellness Letter, 1996, "Why Do Those #&*?@! Experts Keep Changing Their Minds?" February, pp. 4-5.

VanderKam, James C., 1994, *The Dead Sea Scrolls Today*, Eerdmans.

Vaughan, Lewis, 1997, "What's Wrong with Relativism, Why Postmodernism's Most Radical Doctrine Is Dead in the Water," *Free Inquiry*, Summer.

Vertosick, Frank, Jr., 1996, "A Bullet to the Mind," *Discover*, October, pp. 38-40.

Waldrop, M. Mitchell, 1992, *Complexity: The Emerging Science at the Edge of Order and Chaos*, Simon & Schuster.

Wallace, Philip R., 1996, *Paradox Lost: Images of the Quantum*, Springer.

Watson, Lyall, 1995, *Dark Nature, A Natural History of Evil*, HarperCollins.

Weinberg, Steven, 1994, *Dreams of a Final Theory*, Vintage.

Weiner, Jonathan, 1995, "Evolution Made Visible," *Science*, 6 January, pp. 30-33.

Welsch, Roger L., 1994, "Astrophys Ed," *Natural History*, February, pp. 24-25.

Wicher, Enos R., 1974, "God Exists?" *Physics Today*, August, pp. 15, 70.

Wilber, Ken, 1985, *Quantum Questions: Mystical Writings of the World's Great Physicists,* New Science Library.

Wilson, Edward O., 1995, *Naturalist,* Warner.

Witztum, Doron, Eliyahu Rips, and Yoav Rosenberg, 1994, "Equidistant Letter Sequences in the Book of Genesis," *Statistical Science*, Vol. 9, No. 3, pp. 429-438. See also editor's remark on p. 306.

Wolf, Rainer, 1996, "Believing What We See, Hear, and Touch: The Delights and Dangers of Sensory Illusions," *Skeptical Inquirer*, May/June, pp. 23-30.

Wright, Lawrence, 1995, "Double Mystery," *The New Yorker*, 7 August, pp. 45-62.

Wright, Karen, 1997, "When Life Was Odd," *Discover*, March, pp. 52-61.

Wright, Robert, 1990, "The Intelligence Test" (review of *Wonderful Life*, by Stephen Jay Gould), *The New Republic*, 29 January, pp. 28-36.

Young, Matt, 1998a, "The Bible Code: Did God Write the Bible, Or Was It a Computer?" *Rocky Mountain Skeptic*, March/April, pp. 1, 4-6.

Young, Matt, 1998b, "The Bible as a Science Text," *Rocky Mountain Skeptic*, November/December, pp. 2-4.

Young, Matt, 2000, "Judaism as Group Selection," *Skeptic*, Vol. 8, no. 1, p. 21.

Young, Matt, 2001, "Specious Arguments: Twisting Scientific Theory and the Bible," *Skeptical Inquirer*, Vol. 25, No. 2, pp. 51-52 March/April.

Zuker, Charles, 1994, "On the Evolution of Eyes: Would You Like It Simple or Compound?" *Science*, 5 August, pp. 742-743.

Additional bibliography

Armstrong, Karen, 1994, *A History of God*, Ballantine.

Bernstein, Jeremy, 1976, *Einstein,* Penguin.

Bronowski, J., 1965, *Science and Human Values*, Harper & Row.

Brown, Robert Hanbury, 1986, *The Wisdom of Science: Its Relevance to Culture and Religion*, Cambridge University Press.

Churchland, Paul M., 1988, *Matter and Consciousness*, revised ed., MIT Press.

Dyson, Freeman, 1981, *Disturbing the Universe*, Harper Colophon.

Einstein, Albert, 1934, *Essays in Science*, Philosophical Library.

Feinsilver, Alexander, 1980, *The Talmud for Today*, St. Martin's.

Firsching, F. Henry, 1997, *The God Hypothesis, A Scientist Looks at Religion*, Think about It Press.

Gaarder, Jostein, 1996, *Sophie's World, A Novel about the History of Philosophy*, Berkley.

Hempel, Carl G., 1966, *Philosophy of Natural Science*, Prentice-Hall.

Heschel, Abraham Joshua, 1987, *I Asked for Wonder, a Spiritual Anthology*, Crossroad.

Hick, John H., 1973, *Philosophy of Religion*, Prentice-Hall.

Hofstadter, Douglas F., 1980, *Gödel, Escher, Bach: An Eternal Golden Braid*, Vintage.

Jastrow, Robert, 1980, *God and the Astronomers* (new edition with afterword), Warner.

Jastrow, Robert, 1981, *The Enchanted Loom: Mind in the Universe*, Simon and Schuster.

Kemelman, Harry, 1982, *Conversations with Rabbi Small*, Fawcett Crest.

Kushner, Harold S., 1994, *To Life! A Celebration of Jewish Being and Thinking*, Warner.

Kushner, Harold S., 1989, *Who Needs God*, Summit.

Kushner, Harold S., 1987, *When All You've Ever Wanted Isn't Enough: The Search for a Life that Matters*, Pocket Books.

Lederman, Leon, 1993, *The God Particle: If the Universe Is the Answer, What Is the Question?* Houghton Mifflin.

321

Lindsay, Robert Bruce, and Henry Margenau, 1957, *Foundations of Physics*, Dover.

Pagels, Heinz R., 1986, *Perfect Symmetry: The Search of the Beginning of Time,* Bantam/New Age.

Pais, Abraham, 1994, *Einstein Lived Here*, Oxford University Press.

Pelikan, Jaroslav, 1990, *The World Treasury of Modern Religious Thought*, Little, Brown.

Peters, Ted, ed., 1998, *Science and Theology: The New Consonance*, Westview.

Ruse, Michael, ed., 1988, *But Is It Science? The Philosophical Question in the Creation/Evolution Controversy*, Prometheus.

Sacks, Oliver, 1987, *The Man Who Mistook His Wife for a Hat and Other Clinical Tales*, Perennial Library.

Schulweis, Harold M., 1994, *For Those Who Can't Believe: Overcoming the Obstacles to Faith*, HarperCollins.

Wright, Robert, 1988, *Three Scientists and Their Gods*, Times Books.

Index

0.05 level, 148
a posteriori probability, 221
a priori probability, 221
Aaron, 83
Abolitionist, 159
abortion, 90, 175
Abraham, 81, 136, 144, 150
Achilles, 242
achromatopes, 244
action at a distance, 32, 304-307
ad hoc hypotheses, 33, 69, 71, 94, 112, 209
ad hoc hypothesis, 21, 22, 33, 35, 37, 39-42, 68, 69, 81, 85, 196, 294
Adam and Eve, 124
Adonai, 129, 131, 132, 134, 135, 150
Adonis, 147
adversary, 161
afterlife, 97, 235, 240, 243, 289, 291
age of Aquarius, 34
agnosticism, 250, 253
alchemists, 17, 18

alcohol, 28, 78, 151, 152, 292
Alfonso X, 167
aliens, 241
allegory, 88, 117, 126, 203, 286
Allen, Steve, iii, 16, 93, 115, 159, 160, 289
alpha particles, 25-5
altruism, 169, 170, 172-176, 182, 183, 249, 251
Alzheimer's disease, 83
Amalek, 181
American Revolution, 92
Amir, Yigal, 105
anal-erotic, 37
Anderson, Bernhard, 130, 137, 160, 161, 163, 165
anecdotal evidence, 28, 30, 38, 79, 148
anecdote, 12, 30, 75, 76, 78, 79, 193, 238
anecdotes, 10, 30, 31, 73, 79, 80, 239, 245, 271, 272
anomalies, 69, 199, 251

birch trees, 118
birth chart, 34, 36
black hole, 202
black holes, 201, 223, 224
Blackmore, Susan, 96, 97, 235-237
blik, 68, 69
body, 234-239, 254, 258, 259, 263, 304
body in motion, 199
body of Jesus, 139-141
Boeing 747, 62, 63, 221
Bohm, David, 257, 307
Book of Acts, 145
Book of Proverbs, 160
Book of Jonah, 5, 99, 133-137, 272
Book of Numbers, 57, 64, 100, 101, 104, 118, 191, 208, 226, 236, 238
Book of Job, 6, 125, 160-162, 164-169, 273
Book of Genesis, 5, 100, 102, 107, 110, 121, 128, 130, 137, 148, 272
Book of Deuteronomy, 100, 101, 104, 106, 128, 130
Book of Esther, 137

Book of Ecclesiastes, 160, 187
Born, Max, 225, 304
Bower, Bruce, 77
Boyajian, George, 215
bra, 300
brain, 4, 6, 178, 234-240, 254, 255, 259-264, 266, 267, 275, 292
Bravo, Phil, 74, 75
Brecht, Bertold, 282
Bronowski, Jacob, 229, 280
Bronstein, Herbert, iv
Brownian motion, 230
Bruno, Giordano, 231
Buddha, 243
Buddhism, 158, 242
Burgess shales, 211
Bush, Guy, 109, 119, 211, 212
Butler, Samuel, 170

Cambrian explosion, 216, 217
camera eye, 64, 65
Canaan, 159
Carter, Dan, 12
Casti, John, 50
Catholic church, 10, 94
Catholics, 16, 93, 158
causality, 28, 159
causation, 28, 73, 76

cause and effect, 197, 304
cerebrum, 259
cetaceans, 174
chair, 225-228
chance, 100, 101,
    106-108, 147, 148,
    255-257
Chang, Michael, 75
Chassidim, 115
chemistry, 57, 149, 228,
    264, 266, 269
chess, 227, 228, 254, 292,
    297
child abuse, 37
China, 75
Chinese civilization, 99
chordate, 216
Christianity, 93, 94, 138,
    142, 146
Christians, 93, 94, 147,
    158, 240, 243, 246
church, 10, 78, 80, 82, 92,
    94, 98, 143, 151, 159,
    252, 282
circular argument, 222
clairvoyance, 16, 283
cold fusion, 46, 47
colorblindness, 244
Columbus, Christopher,
    11
Communism, 93
community of shared
    values, 14

complexity, 203, 210,
    212-215, 218, 252,
    274, 293
Conan Doyle, Sir Arthur,
    242
consciousness, 168, 241,
    267, 303
constructs, 13, 250
contiguity, 304, 305
continental drift, 52, 58
contingent, 204-206
control group, 39, 40,
    260, 261, 292
convection cells, 202
conventional wisdom, 20
convergence zones, 259
cooperation, 172, 173,
    175, 179
Copernican revolution,
    48, 51, 52
Copernicus, Nicholas, 51,
    52, 218, 231
Copleston, F. C., 199,
    200
correlation, 28, 30, 105,
    107, 121, 159
correlations, 101, 103,
    106-108, 111, 112, 307
cosmic time, 121
cosmology, 4, 6
Coughlin, Father, 157
covariance, 29
covenant, 246

339

342

About the Author

Matt Young is Adjunct Professor in the Department of Physics and the Division of Engineering at the Colorado School of Mines in Golden, Colorado, and was formerly Physicist with the National Institute of Standards and Technology in Boulder, Colorado. While with NIST, he earned the Department of Commerce Gold and Silver Medals for his work in optical fiber communications and was named Fellow of the Optical Society of America. His previous positions include Assistant Professor of Physics at the University of Waterloo, Assistant Professor of Electrophysics and Electronic Engineering at Rensselaer Polytechnic Institute, Associate Professor of Natural Science at Verrazzano College, and Visiting Scientist at the Weizmann Institute of Science.

Professor Young is a member of the Optical Society of America, the American Association for the Advancement of Science, the Federation of American Scientists, the Committee for the Scientific Investigation of Claims of the Paranormal, and the Rocky Mountain Skeptics. He is the author or co-author of roughly 100 publications and was twice Guest Editor of the international Conference on Precision Electromagnetic Measurements. He is the author of *Optics and Lasers, including Fibers and Optical Waveguides* (fifth edition, 2000) and *The Technical Writer's Handbook: Writing with Style and Clarity* (1989); both books have appeared in foreign translations and are still in print.

Finally, Professor Young is a former Trustee of Congregation Har HaShem, a Reform synagogue in Boulder, Colorado, and of the Hillel Council of Colorado, a Jewish campus organization.

349